JEFFERSON *and* SCIENCE

JEFFERSON *and* SCIENCE

by Silvio A. Bedini

Preface by Donald Fleming

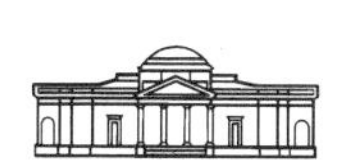

THOMAS JEFFERSON FOUNDATION

Monticello Monograph Series

2002

Library of Congress Cataloging-in-Publication Data

Bedini, Silvio A.
Jefferson and science / by Silvio A. Bedini.
p. cm. -- (Monticello monograph series)
Includes bibliographical references and index.
ISBN 1-882886-19-4
1. Jefferson, Thomas, 1743-1826--Contributions in science. 2. Science--United States--History--18th century. 3. Science--United States--History--19th century. 4. Presidents--United States--Biography. 5. Scientists--United States--Biography. I. Title. II. Series.

E332.2 .B365 2002
973.4'6'092--dc21

2002035858

On the Cover

Inset images: Jones-type improved compound microscope with slides, similar to one purchased by Jefferson on April 12, 1786, from John Jones, along with a botanical hand magnifier owned by Jefferson (Thomas Jefferson Foundation, Inc.). Below, jawbone of American mastodon (courtesy of the University of Virginia). Background: cipher devised by Jefferson for confidential communications with Lewis and Clark on their expedition (courtesy of the Library of Congress).

Unless otherwise noted, Edward Owen is the photographer for images owned by the Thomas Jefferson Foundation and the University of Virginia.

Designed by Gibson Design Associates.
Edited and coordinated by Beth L. Cheuk; editorial assistance provided by Christine McDonald.

This book was made possible by support from the
Martin S. and Luella Davis Publications Endowment.

Distributed by
The University of North Carolina Press
Chapel Hill, North Carolina 27515-2288
1-800-848-6224

Nature intended me for the tranquil pursuits of science, by rendering them my supreme delight. But the enormities of the times in which I have lived, have forced me to take a part in resisting them, and to commit myself on the boisterous ocean of political passions.

—Jefferson to Pierre Samuel
Du Pont de Nemours,
March 2, 1809[1]

Thomas Jefferson Foundation, Inc.

CONTENTS

PREFACE

Anyone who reads Silvio Bedini's admirable survey is bound to find the sheer breadth of Thomas Jefferson's horizons almost inconceivable to people of our time. The challenge becomes to grasp his intellectual location in history and conceptualize him among his contemporaries and successors.

A first approximation to locating Jefferson both historically and functionally is to identify him as a specimen of the would-be "universal man" of the eighteenth century, endeavoring to encompass the broadest possible span of human concerns. This was a rapidly vanishing, though not quite extinct, aspiration as the learned increasingly preferred their excavations to be deep and narrow rather than wide and shallow. The prime indication of this was the unending segmentation of science into new forms of specialization with elaborate modes of certification attached. This increasingly entailed consigning "unqualified" persons to the pejorative category of mere "amateurs," from which Jefferson escaped.

Of Jefferson's contemporaries in Europe, Goethe was virtually a universe of his own whose scope extended to christening and helping to found the science of morphology and, less happily, to rejecting Newton's theory of colors. But the quintessential representatives of the universal man in Jefferson's time, and perceived by Jefferson and others as valedictorians of the breed, were Wilhelm (1767-1835) and Alexander (1769-1859) von Humboldt, Prussian-born, but men of the world both literally and metaphorically.

Wilhelm von Humboldt was the author of a major treatise advocating the minimal state (quoted on the title-page of John Stuart Mill's *On Liberty*); a Prussian diplomat; minister to London, Rome, and Vienna and one of the main negotiators at the Congress of Vienna; Prussian Cabinet minister with substantial input to a new constitution for Prussia; and the principal founder of the University of Berlin (now named after him). He was a polylinguist with a specific interest in the Amerindian languages. More fundamentally, he was one of the chief pioneers of modern linguistics.

When Wilhelm von Humboldt was dismissed from public service, he retired to Tegel, his family home in Prussia. Here he supervised its total remodeling and enlargement, accompanied by the creation of magnificent gardens. All of this was executed by the greatest German architect of the nineteenth century, Karl Friedrich Schinkel, whom Humboldt had discovered. For Wilhelm von Humboldt and later Germans, Tegel was the exact analogue to Jefferson's Monticello—the most intensely personal expression of their abounding creativity. Monticello survives. Tegel was gutted by Russian soldiers in 1945.

The younger Humboldt, Alexander, was a mining engineer and geologist by training, the inventor of a safety lamp and rescue apparatus for miners, a contributor to chemistry, and an incessant agitator for a worldwide network of geodetic stations and astronomical observatories. He became one of the greatest of all scientific explorers in a five-year expedition to Latin America from 1799 to 1804, a subsequent explorer of Russian Asia, a pioneer climatologist who invented isotherms and isobars, the discoverer of the pivotal role of volcanic activity in geological history, and the founder of the science of phytogeography. More generally, he is often seen as one of the main precursors of ecology.

Alexander von Humboldt was an acute observer of Latin American Indians and differentiator of their various cultures, and accordingly a major contributor to ethnology. He scoffed at the characterization of Amerindians as "barbarous." In this as in other respects, Alexander von Humboldt's massive account of his Latin American findings resembles the main thrust of Thomas Jefferson's much earlier *Notes on the State of Virginia* (initial private publication 1785). Both works constitute emphatic refutations of the notorious claim—forwarded by various European writers who had never been to America—that the enervating physical environment of the New World inevitably bred inferior animals and human beings. This was the very claim that led Jefferson to tell Lewis and Clark to be on the lookout for larger-than-European animals (or at least their bones). In the same cause, Jefferson, who never lived among Indians in their native habitats as distinguished from receiving chieftains on embassies to the white man, invoked a prime example of the Noble Savage in the person of "Logan, a Mingo chief." Humboldt, who had lived among Indians with only one European companion for years on end, was decidedly nuanced in his praises of them. But it was Jefferson who described them as "a barbarous people." Notwithstanding this, both men were vindicators of the New World and its inhabitants against their European detractors. Symbolically, Alexander von Humboldt became the first to propose a Panama Canal, a friend and adviser to Simon Bolivar, and an enthusiast for independent republics in Latin America.

Jefferson's *Notes on Virginia*, barely of book length, remained his only extended disclosure of his scientific interests. It was utterly dwarfed by Alexander von Humboldt's gigantic uncompleted master-work *Kosmos*, intended to cover the whole earth. Apart from its brevity, Jefferson's *Notes* had been confined to North American issues. Yet the two men had classically illustrated, on their respective scales, the universalizing impulse authorized by their own times but alien to ourselves.

The parallel is sharper than this. Between them, the Humboldt brothers spanned almost exactly the entire range of Jefferson's specific intellectual concerns, with the sole exception of religion—nation-founding or reconstituting, political theory, instrument construction, scientifically oriented exploration, metrology (the science of measurement), astronomy, geology, geodetics, physical geography, botany, zoology, meteorology, climatology, ethnology, racial conceptions, linguistics, architecture, and garden design.

No imitation from either side was involved. Yet it was fitting that shortly before he returned to Europe in 1804, Alexander von Humboldt, on paying his respects to the president of the United States was invited by Jefferson, who was already well informed about his exploration, to visit Monticello, where Humboldt ended up staying for three weeks. The two men corresponded thereafter, and Humboldt regularly dispatched new fascicles of his voluminous publications on South America. At Humboldt's request Jefferson sent *him* a copy of *Notes on Virginia* with the disclaimer that it would seem negligible "to the author of the great work on Latin America." These exchanges constituted an emblematic mutual recognition of their intellectual kinship as specimens of the universal man, though of unequal stature in this role.

This emphatically did not mean that they enacted the role identically. Alexander von Humboldt was a private man of independent means who financed his own expeditions. He put his queries to nature in person—supremely so in his five years of confronting the Latin American jungle. In sharp contrast, Jefferson became a head of state able to dispatch Lewis and Clark to interrogate nature on his (and the nation's) behalf. Yet the questions were his own and only a person of his broad knowledge could or would have framed them. Besides, he himself conducted researches closer to home.

By no accident whatever, the only comparable public figures in American history are Jefferson's respectively older and younger contemporaries, Benjamin Franklin and John Quincy Adams. But Franklin by his major contributions to founding the science of electricity—arguably more important than any single achievement by Alexander von Humboldt—transcends the comparison though otherwise conforming to the same general pattern.

John Quincy Adams, no scientist himself, qualifies for inclusion by his famous treatise on the reform of weights and measures (also one of Jefferson's concerns), his virtually religious consecration to promoting astronomical observatories in America, and his valiant struggle against the odds to get the Smithson bequest devoted to scientific research rather than college education or patronage jobs. Adams's nearest approach to expertise about science, as in a famous oration of 1843, lay in chronicling the history of astronomy through the ages to encourage his fellow-Americans to do their share in turn. He knew from his study of weights and measures, in a report of 1821 as secretary of state, that astronomy had practical uses and required different vantage points. But the emotional thrust of his exhortations was to remind humanity in common that they been given "a heavenward looking face," as if "the special purpose of their creation" had been "the observation of the stars."

Jefferson too wanted to enlist the entire world in fostering science, but he would never have invoked edifying astrotheology for the purpose. The predominant bent of his own far ampler embodiment of the universal man was unabashedly practical, as in his passion for new and improved scientific instruments and mechanical devices. Yet his practicality must also be understood to include acquiring useful knowledge for the conduct of statecraft, as in framing the agenda for the Lewis and Clark expedition, though indubitably accompanied as always by sheer curiosity. Here as elsewhere, Jefferson harnessed his curiosity to national purposes. This was inevitable in one of the founders of a nation and its subsequent leader. It yielded the paradox, though not the contradiction, of a universal man who put his learning to the service of an enlightened nationalism. Americans would never look upon his like again. Even giants require a favoring environment.

—Donald Fleming
Harvard University

Thomas Jefferson *by Cornelius Tietbout after Rembrandt Peale, 1801. At Jefferson's left are an electrostatic machine and a terrestrial globe (Thomas Jefferson Foundation, Inc.)*

INTRODUCTION

When your mind shall be well improved with science, nothing will be necessary to place you in the highest points of view but to pursue the interests of your country, the interests of your friends, and your own interests also with the purest integrity, the most chaste honour.

—JEFFERSON TO PETER CARR, 1785[2]

Thomas Jefferson was born with a mind that was never still, and a constant and insatiable curiosity. Born and raised at the edge of the Virginia frontier, he was inspired from boyhood on by the surrounding world of nature and the unknown that lay beyond the beyond. The son of a successful pioneer and surveyor, he learned the importance of the practical sciences at an early age, which he applied often on behalf of the developing American colonies and an emerging new republic. At the same time that he encountered the limiting restraints of frontier life, he learned to appreciate the beauties as well as the hazards that lay in the wilderness. He also developed an abiding interest in its flora and fauna, along with a concern for the land, its development, and the natural world. The latter was likely kindled by his father, who also taught him how to write, introduced him to the magic of mathematics, and instructed him in the elements of surveying.[3]

Jefferson's interest in the sciences was fueled not only by learning about his father's surveying experiences but also by several of his teachers. At the Reverend James Maury's school he first saw natural curiosities including fossils and geological specimens, and he may have been at liberty to read in the schoolmaster's extensive library. Among the authors he encountered in his student years were Cicero, Epictetus, and Shakespeare, and he was particularly impressed by the writings of Francis Bacon, Isaac Newton, and John Locke. As he wrote to Dr. Benjamin Rush in 1811, "Bacon, Newton and Locke … were my trinity of the three greatest men the world had ever produced."[4]

William Small (courtesy of The College of William and Mary)

There is no doubt that the strongest influence on his education, the one who directed him to scientific pursuits, was Dr. William Small, his professor of natural and moral philosophy, with whom he studied at the College of William and Mary. During his college years, young Jefferson became Small's privileged associate as they shared their hours of leisure.[5]

During this period, Jefferson also developed friendships with Virginia Governor Francis Fauquier and the lawyer George Wythe, who with Small formed a small intellectual circle to which Jefferson was admitted. Men of wide interests, they participated in musical concerts in the Governor's Palace and engaged in lively discussions on questions of natural philosophy as well as events of the times. Fauquier's account of an unusual hailstorm was reported to the Royal Society of London, and his weather diary, maintained from 1760 to 1762, was subsequently published. Undoubtedly Fauquier's example inspired Jefferson's lifelong system of weather observation.[6]

Closely associated with Jefferson's scientific interests was the young man's compulsion to collect and record in pocket memorandum books random bits of information, which eventually he might find useful and would apply to practical purposes as opportunities arose.[7] Included were most unlikely combinations of data, some useful and some trivial, often related to measurement in one form or another. Entries ranged from the weight of a dwarf at birth, the number of cotton seeds required to fill a bushel, or how much land that bushel would plant if the seeds were placed four to a hill spaced two feet apart. He justified his preoccupation with apparent miscellaneous minutiae in a statement in his *Notes on the State of Virginia* that "a patient pursuit of facts, and cautious combination and comparison of them, is the drudgery to which man is subjected by his Maker, if he wishes to attain sure knowledge."[8]

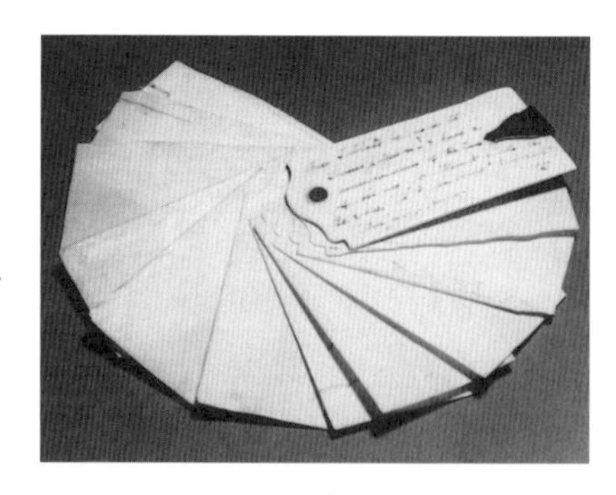

Portable and erasable ivory notebook (Thomas Jefferson Foundation, Inc.)

Chapter One

SURVEYING

I have found this, in practice, the quickest and most correct method of platting which I have ever tried, and the neatest also, because it disfigures the paper with the fewest unnecessary lines.

—JEFFERSON TO LOUIS HUE GIRARDIN, 1814[9]

Young Jefferson became familiar with surveying at an early age, for his father, Peter Jefferson, was a prominent surveyor of the Virginia Piedmont, much of which was unexplored. Peter Jefferson and his friend and fellow surveyor Joshua Fry, professor of mathematics at the College of William and Mary, often undertook surveying expeditions together. They completed a map of Virginia, published in 1751, that was of the greatest importance. This work required long absences from home and family, and the exciting tales of adventure and danger that they told on their return intrigued Jefferson.[10]

Jefferson's father taught his son surveying and bequeathed him his instruments. From time to time Jefferson purchased other instruments, the earliest of which he acquired in January 1778 while pursuing his law career with George Wythe in Williamsburg. It was "a common Theodolite or graphometer 8 I[nches] 54 D[ollars]," indicating that it measured eight inches in diameter and had been purchased for fifty-four dollars. It may have been the same instrument noted in his records for which he had paid Reverend Andrews the sum of forty-five pounds. This he used for making astronomical observations in November of the same year.[11]

He reported how after "placing the Theodolite on the top of the house [he observed] the Eastern spur of the High Mountain intersects the Horizon 90°

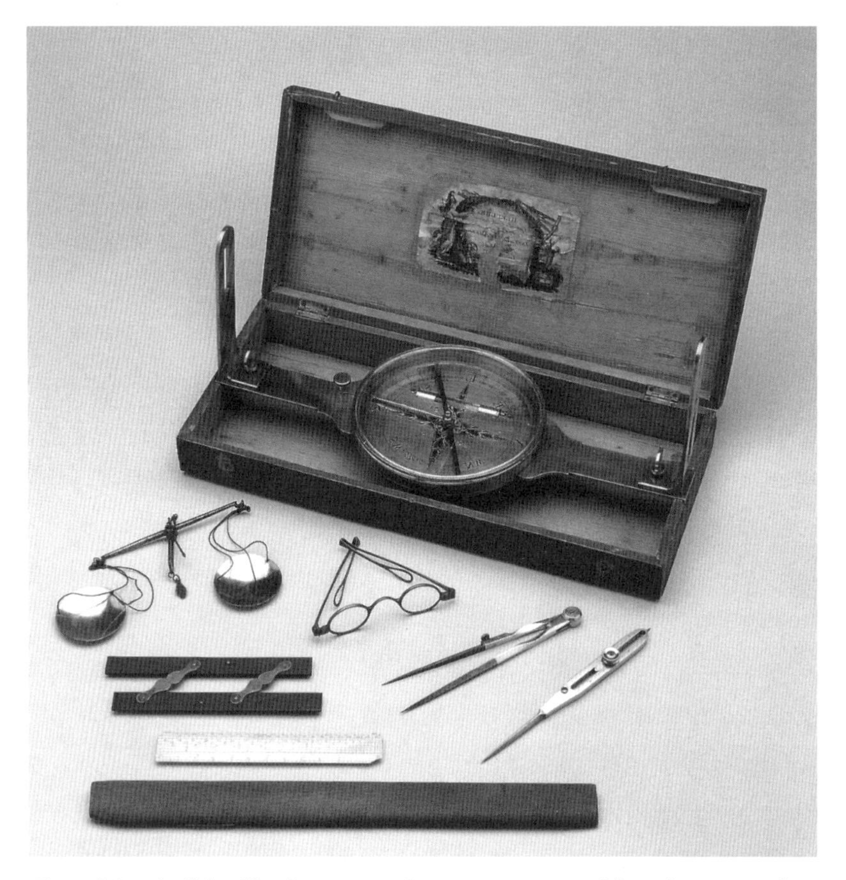

From Monticello's collection: surveying compass, portable scales, spectacles, and drafting instruments (Thomas Jefferson Foundation, Inc.)

Westward of Willis's mountain. Note the observation was made on the intersection of the ground (not the trees) with the horizon." Inasmuch as Willis Mountain was the isolated peak easily visible from Monticello on clear days, Jefferson used it again and again as a focal point for calculating the latitude and the longitude of various points in the region.[12]

In 1773, when Jefferson was thirty and had studied law with George Wythe, the president and faculty of the College of William and Mary "unanimously agreed that Mr. Jefferson be appointed surveyor of Albemarle [County], in the room of Mr. Nicholas Lewis, who has sent his letter of resignation, and that he be allowed to have a deputy." The appointment came at a crucial time when Jefferson was deeply depressed by the state of his law practice and was contemplating seeking another profession, not yet chosen. Although surveying would not have been his choice for a lifetime career, it was nonetheless an opportunity, and he decided to make use of it. It was evident that he prepared for the county position by purchasing "Marshall's meridian instrument" for twenty shillings in Virginia currency.[13]

Marshall's instrument provided a welcome solution to Virginia surveyors faced with the problem of magnetic declination. The maker of the instrument was not identified, but no professional makers of instruments are known to have worked in Virginia in this period, so it must have been manufactured elsewhere, possibly in Philadelphia. Despite the most exhaustive search over a period of years, no example of the instrument has come to light, and the only known record of purchase was the entry "Marshall's meridian instrument ma[hogany] cur/20#" made by Jefferson in his undated inventory of his scientific instruments compiled in 1784.[14]

There is no evidence that Jefferson actually engaged in surveying during the brief period of his appointment as county surveyor; no surveys signed by him were returned and filed. Apparently he allowed his appointment to lapse the following spring, and he would have had several reasons for doing so. After accepting the appointment, he probably realized upon reconsideration that most of the desirable land in the county had already been assigned, and great profit was no longer to be made thereafter from surveying in the field. Recalling his father's way of life, he realized that he would have to be absent from home for extended periods of time without possibility of communication in the interim.

Theodolite by Jesse Ramsden believed to be purchased by Jefferson in 1778 (Thomas Jefferson Foundation, Inc.)

In 1777 Jefferson returned to professional surveying when he surveyed the entire Albemarle County, in which he was living, for the purpose of establishing the area to be assigned for a newly created county to be named Fluvanna. He produced a map of Albemarle County on which he defined the proposed division, indicated by means of a line running from the southwest corner of Louisa County to the downstream side of Scotts Ferry on the Fluvanna River. In addition, Jefferson made surveys of his own properties at Monticello and Poplar Forest and for some neighbors.[15]

Jefferson had a particular love for scientific instruments, and he sought and purchased examples of the finest surveying and astronomical instruments available. During his sojourn in England he visited the shops of the most prominent makers of mathematical instruments and made numerous purchases. Among these was an excellent theodolite made by the well-known London maker Jesse Ramsden. The instrument, which survives, measures both horizontal and vertical angles by means of a movable upper telescope and a fixed one below the plate. The vertical arc and the toothed rack enable the upper telescope to be adjusted precisely to any elevation, while the upper surface of the flat horizontal plate can be rotated simultaneously to any position around a full 360 degrees. A comparison can then be made between the two telescopes. Jefferson was fond of the instrument and used it frequently for surveying local land areas and for determining the heights of neighboring mountains.[16]

The surveying instrument of which Jefferson was the most proud, however, was one he acquired late in his career. It was unquestionably one of the most sophisticated scientific instruments of the time—a Universal Equatorial Instrument made by Jesse Ramsden. He purchased it second-hand in 1792 from John William Gerard De Brahm, the German military surveyor and engineer of Georgia. The instrument was made portable by means of an elaborate equatorial mounting, based upon the principle of the helioscope, which could be set by clockwork to follow the course across the sky of an observed celestial body and thus provide a continuous record. The one acquired by Jefferson appears to have been the only example owned and used on the American continent. After experimenting with it for some time, Jefferson devised an improvement consisting of the addition of a twelve-inch telescope to be fitted to the main instrument, which he had made for him by the London maker William Jones.[17]

Among Jefferson's other surveying equipment was a theodolite made by W. & T. Gilbert of London, a firm in existence between 1810 and 1842, and acquired during Jefferson's later years. Its damaged condition suggests that it had been stored in the Rotunda of the University of Virginia and suffered damage during the fire of 1895 that destroyed the building.[18]

The distances he traveled fascinated Jefferson, and he kept copious notes of his journeys. An instrument that he found most useful and intriguing was the odometer, a device that measured the distance traveled by counting and recording the revolutions of a wheel, such as a carriage wheel of specified circumference. He experimented with various types of the instrument, and on several occasions while he was in France, he attempted to purchase odometers from London. Upon learning that the artist John Trumbull was about to depart for England, Jefferson asked him to inquire there concerning a triangular odometer about which he had been told. Shortly before his return to the United States, Jefferson sought information from Benjamin Vaughan, a British diplomat in London, about another odometer, writing, "They make in London an Odometer, which may be made fast between the two spokes of any wheel, and will indicate the revolutions of the wheel by means of a pendulum which always keeps it's vertical position while the wheel is turning round and round."[19] Jefferson asked Vaughan to obtain what information he could about the dependability of the instrument and its cost, but he did not manage to acquire one. In 1791, after his return to the United States, Jefferson purchased an odometer from the Philadelphia clockmaker Robert Leslie and three years later acquired another from David Rittenhouse.[20]

In 1791, while serving as secretary of state, Jefferson utilized an odometer for the first time to measure the distance from Washington to Monticello, with the instrument attached to the wheel of his phaeton. He noted

> *These measures were on the belief that the wheel of the Phaeton made exactly 360 revolutions in a mile. But on measuring it accurately at the end of the journey it's circumference was 14 ft. 10-1/2 I[nches] and consequently made 354.95 revol[utions] in a mile. These numbers should be greater then in the proportion of 71:72 of a mile added to every 71.*[21]

Later Jefferson ordered odometers constructed to his own designs and specifications, and he became particularly involved in this endeavor with James Clarke, a landowner, local politician, and inventor living in Powhatan, Virginia. Their association continued to almost the end of the statesman's life. Jefferson first learned of Clarke in the spring of 1807, hearing that Clarke "had invented and made a machine to be fixed behind a carriage for counting the revolutions of the wheel while travelling." It appeared to be of a type with which Jefferson had experimented, as he informed Clarke, that he "made an effort of the same kind & failed," and inquired whether Clarke would be willing to permit his instrument to be copied.[22]

Clarke not only responded in the affirmative but also offered his invention as a gift that he would bring to Jefferson either in Washington or at Monticello so that he could show the president how to install and disassemble it. Clarke explained that his first version of the device was not equipped with a bell signal and featured a single hand, or index, that revolved once every ten miles. He then developed a second version with two indices and a bell—one index denoting the number of miles traveled, and the other reporting the fractions of miles divided in decimals. Upon reaching the end of each mile a bell rang. The improved instrument was contained within a small brass box "the size of a shaving case," he reported. It was calculated for use on a wheel that was five feet and one inch in diameter, so that in order to use the device, Jefferson would need to replace the rear wheels of his carriage with others of the designated size.[23]

After becoming acquainted with Clarke, Jefferson commissioned him to produce an odometer of unusual form. Jefferson later described it to a correspondent in a discussion of the various proposals that had been made to the federal government for its establishment of standards of weights, measures, and coinage. Jefferson suggested that each standard should be divisible decimally at the will of the individual. As an example of its popular application, he mentioned one means by which he had tested its applicability. He wrote

> *I have lately had a proof how familiar this division into dimes, cents and mills is to the people when transferred from their money to anything else. I have an odometer fixed to my carriage which gives the distances in miles,*

> *dimes, and cents. The people on the road inquire with curiosity what exact distance I have found from such a place to such a place; I answer, so many miles, so many cents. I find they universally and at once form a perfect idea of the relation of the cent to the mile as a unit.*[24]

Jefferson felt assured that they would do the same with yards of cloth, pounds of shot, and ounces of silver or medicine, as examples. He was especially proud of this device, and mentioned it also in his autobiography, stating, "I use when I travel, an odometer of Clarke's invention, which divides the mile into cents, and I find everyone comprehends a distance readily, when stated to him in miles and cents."[25]

Jefferson made considerable use of Clarke's odometer, and a decade later he again felt the need to communicate with the inventor. He reported that recently he had been experimenting with the design of another "invention" on the principle of the odometer to be used for surveying. He described the instrument and its intended function as "a machine that could lay down the platt of a road by the traveling of a carriage over it." After a number of unsuccessful attempts to make it operable, he finally put it aside.[26]

Clarke promised to take up the challenge and determine what could be done with it. Eventually he succeeded in producing a device that fulfilled the purpose that Jefferson had intended, to be attached to a carriage wheel to record the direction taken.[27] As Clarke explained,

> *The chart is placed on a horizontal plane on the floor of the carriage. The part of the machine which marks off the platt, progresses on the chart as the carriage moves. A rod, or index, placed convenient to the eye, and moved by the hand, is kept constantly pointing to the North Star. The index being connected with the machine, and changing its angle with the carriage, at every turn of the road, produces on the chart all the turns and angles which there is on the road.*
>
> *But, altho this machine is perfect in principle, it is not so in practice; for want of a perfect index. The objections to this index are. 1st. It*

can be used only in the night. 2d. It can be used only in fair weather. 3d. It can be used only in latitudes, neither too high nor too low. 4th. The North Star is too often eclipsed by trees and other objects; particularly in summer.

I have tried several experiments to make the magnetick needle answer as an index; but have not been able to succeed; in consequence of the vibration of the needle produced by the agitation of the carriage. If this difficulty could be removed, I have no doubt it would be a very valuable acquisition. And what would add still more to its convenience—the ascent, and descent, of unlevel ground could be taken upon the same principle, and at the same time.

To take the ascent & descent, a pendulum would be a very good index. Altho the pendulum would be constantly vibrating, its general direction would be perpendicular to the horizon, and as the plane of the chart varies from the plane of the horizon, by the ascending or descending of the carriage, so will the pendulum vary from its right angle with the plane of the chart, and mark off a trail thereon, above, or below, a right line on the chart, representing the horizon.

In surveying water courses, navigable by boat, or even canoes, the magnetic needle would answer exceedingly well as an index to this machine if the water is still; but as the machine would receive its motion from the surface of the water, the current would defeat the object, and render it impracticable.

In surveying public roads with this instrument, the magnetic needle will answer very well, by the addition of one person more, to go before with a white pole, and stopping the carriage at each turn of the road, long enough for the needle to rest, and to set the index.

And this I think will be a great improvement on the common way of surveying roads; as it will be more expeditious, less expensive, and less subject to error: as the surveyor will not have to keep a reckoning—to enter his notes—and to plott from those notes, all of which are subject [to] error.[28]

As might be anticipated, Jefferson responded with enthusiasm. "I have but occasionally looked at the subject as a desideratum," he replied, "but never seriously aiming at its solution myself." He went on,

> *The basis however of what has occurred to me is a four wheeled carriage, very light, the wheels to be like cotton spinning wheels & all other parts proportionately light. Just over the bolt which connects the perch with the fore axle, suppose a machine fixed, so as to remain steadily in the direction of the perch, and the fore-axle made to govern a tracer which should draw on paper all the changes of angle and direction which the fore axle should commence. But as to the wheelwork & other contrivances necessary to effect this on paper, I never aimed at them. They are much more within your competence. I am afraid you will find the magnetic needle too weak and tremulous an agent to fulfill your views.*[29]

There the matter apparently ended, for there was no further communication on the particular subject between Jefferson and Clarke. When Clarke again wrote to Jefferson, it was in relation to the standard type of odometer Clarke had produced in 1817. He wrote,

> *By the advice and persuasion of several Gentlemen who are anxious to get an Odometer like mine, I have at length concluded to take a patent, and establish a manufactory of them.*
>
> *As you have had one over many years, I'll thank you for your opinion of them, as to accuracy, simplicity, and durability; whether it incommodes, or disfigures a carriage. And whether you believe the plan to be original, or whether you ever saw, or heard of one in Europe or America upon the same plan, previous to the adoption of mine.*[30]

Jefferson responded, assuring Clarke that his odometer was as simple as one could expect such a machine to be, "having only three toothed wheels, entirely accurate, inconsiderable in weight and volume, and of convenient application to the carriage." As to originality, Jefferson added that he knew of no odometer in Europe or America that resembled Clarke's to any degree or that could be compared with it. He still continued to use the one that he had, he noted, "finding great satisfaction in having miles announced by the bell as by milestones on the road."[31]

In the following year Jefferson's odometer was damaged, and again he had recourse to the inventor. He explained how, when returning from his retreat at Poplar Forest he lost the rod and ratchet wheel connecting motion from the carriage wheel to the odometer. Since he could recall none of the details of the wheel's form and size, he asked Clarke to make a diagram on paper for him providing the diameter of the wheel as well as of the rod on which it was put so that his smith at Monticello could make one. Or, perhaps Clarke could send the little wheel and save him the trouble of having to make one. Considerable time passed

Jefferson's odometer made by Nairne & Blunt of London (Thomas Jefferson Foundation, Inc.)

before Clarke received Jefferson's letter but then he lost no time in forwarding "a case containing the rod with the wheels (from the same moulds) already fixed."[32]

Jefferson's preoccupation with odometers was fairly well known, and in 1820 another Virginia inventor, James Deneale of Dumfries, asked the retired statesman for his evaluation of a "land mapper" that he had invented. As the name suggests, it was a device for mapping lands, somewhat similar to that which Jefferson had proposed to Clarke. Deneale explained that he had sent his land mapper for evaluation to someone in Fredericksburg who returned it with the comment that he was unable to believe that it was a new invention. He added that he had been told that Jefferson already used something similar in mapping fields. Deneale then asked Jefferson whether his mapping instrument was similar to his own. Jefferson assured him that there was no similarity, and a month later, in August 1820, Deneale was granted a patent for his invention of a "Mapping Land Instrument."[33]

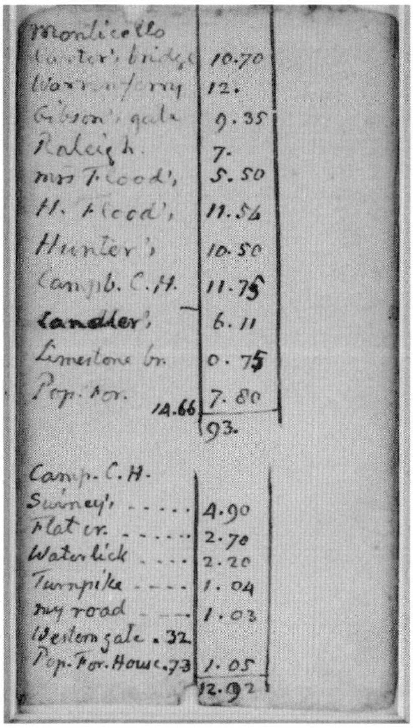

Notes on distances traveled maintained by Jefferson in a pocket record book (courtesy of Special Collections, University of Virginia)

Although land mapping devices made by both Clarke and Deneale had been patented, and appear to have been ingenious and useful advances in the art of surveying and mapping, nothing more is known of them, neither descriptions nor examples known. Nor does anything survive of the odometer made by Clark for Jefferson. Of the substantial number of odometers of various types owned by Jefferson, only one associated with him has survived. He acquired it at an undetermined date; its purchase was not recorded either in his Memorandum Books or in the inventory of his instruments that he compiled in the 1780s. The odometer is of an extremely sophisticated form, made by the prominent London firm of Nairne & Blunt. Its brass movement is housed in a fine walnut case having a hinged front panel. It probably was designed to be attached to a wayweiser or wheelbarrow-like frame, with the drive mechanism connected to the wheel through the open bottom of the case. Reading from the outermost circle of the engraved silvered dial, it indicates the distance traveled in chains, links, yards, poles, miles, and furlongs. It also is provided with a conversion table. No similar example is known, suggesting that it may have been made to Jefferson's specifications from his design.[34]

Chapter Two

METEOROLOGY

Of all the departments of science no one seems to have been less advanced for the last hundred years than that of meteorology.

—JEFFERSON TO GEORGE F. HOPKINS, 1822[35]

Jefferson was among the first in the American colonies to conduct systematic meteorological studies, maintaining records of temperature, rainfall, winds, and other climatological data from his student days to his final years. He believed that only by means of simultaneous observations made at considerable distances could knowledge of the weather be successfully derived. He urged others to keep records similar to his own and in time developed a network of weather watchers among his friends and associates. Jefferson began maintaining weather records while studying law at Williamsburg, possibly inspired by the example of Governor Francis Fauquier. While attending the Continental Congress in Philadelphia, he began to record the weather in a comprehensive notebook style. He noted the temperatures for the first few days of July 1776, indicating that July 4, the day that the Declaration of Independence was accepted by the Congress, was a moderately warm day, although later writers erroneously dramatized the day as one of sweltering heat. The next day he purchased a barometer at Sparhawk's stationery store.[36]

Several months later, in September, he took barometrical readings at various locations based upon *Nettleton's Table*, which he had found in the *Philosophical Transactions* for the year 1725; the Table was part of an article on barometrical observations "at different Elevations above the Surface of the Earth" made by a certain Dr. Thomas Nettleton of Halifax.[37]

Jefferson shared his enthusiasm for meteorological studies with others, particularly with the Reverend James Madison of Williamsburg, and the two developed what constituted a weather service that lasted many years, with comparable records maintained in the mountains at Monticello and the lowlands of Williamsburg.

Francis Fauquier (courtesy of Colonial Williamsburg Foundation)

The greatest difficulty Jefferson and his fellow weather observers experienced in obtaining accurate weather observations was acquiring suitable instruments because scientific glass and optical instruments were not yet being made in the United States. Until the first quarter of the nineteenth century they were imported, chiefly from England, where skilled Italian artisans who had migrated and established themselves there had developed an important industry in the manufacture of thermometers and barometers. When Jefferson sent a record of his recent weather observations together with a thermometer to the iron master Isaac Zane, he noted that the instrument was "the only one to be had in Philadelphia."[38]

At Jefferson's request, Madison had borrowed a thermometer "which appears very sensible as to Heat or Cold, tho' it is so constructed that I cannot ascertain the Accuracy of the Division by plunging it in boiling water; This appears of Consequence especially when we keep correspondent Observations," he reported. This was important because Jefferson and Madison were to make simultaneous observations at Charlottesville and Williamsburg. To test his thermometer for the greatest degree of cold, Madison exposed it to the night air in an open window. During the day he exposed it as much as possible in the open air.[39]

Jefferson recruited as many of his friends and associates as he could for his weather studies. In addition to Reverend Madison and Isaac Zane, participants included Benjamin Vaughan in London, William Dunbar in Mississippi, and Hugh Williamson in Quebec. Additionally, future president James Madison was eventually persuaded to join, charged with the following instructions:

I wish you would keep a diary under the following heads or columns, 1. day of the month. 2. thermometer at sunrise. 3. barometer at sunrise. 6. thermom. at 4. P.M. 7. barometer at 4. P.M. 4. direction of wind at sunrise. 8. direction of wind at 4. P.M. 5. the weather viz. rain, snow, fair at sunrise, &c. 9. weather at 4. P.M. 10. shooting or falling of the leaves of trees, of flours, and other remarkeable plants. 11. appearance or disappearance of birds, their emigrations &c. 12. Miscellanea. It will be an amusement to you and may become useful. I do not know whether you have a thermometer or barometer. If you have not, those columns will be unfilled till you can supply yourself.[40]

Among Jefferson's prized possessions was a thermometer designed for travelers that enabled him to continue to conduct climatological observations even while away from home. Made by Dollond, the London instrument maker, it was a little larger than a mechanical pencil or fountain pen and fitted any pocket with ease. It was the first such instrument that the statesman could use during travel. He may have purchased it at Dollond's shop during his English sojourn, but it is more likely that it had been a gift from Payne Todd in 1816. He had gracefully acknowledged "the handsome keep-sake which has amused me much, and not the less by the puzzle it has afforded me, to find out the method of rectifying it. I at length discovered it, and that it was only necessary to loosen a little a single screw to throw it out of geer and to throw it again after setting the index, it was exactly 10 [degrees] wrong."[41]

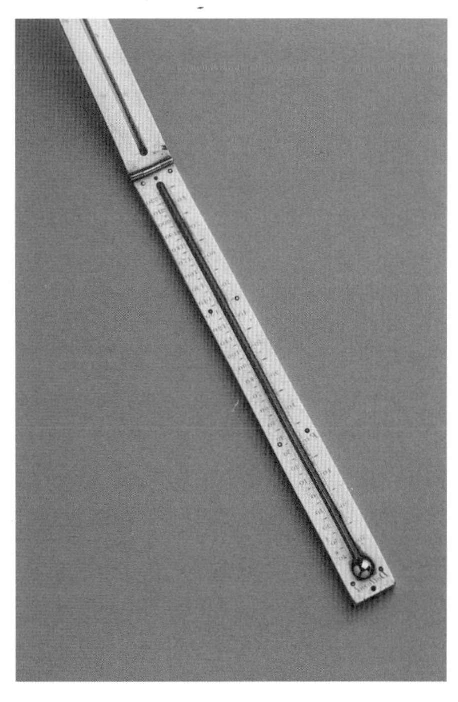

Jefferson's portable brass thermometer made by Peter and John Dollond (courtesy of the National Museum of American History, Smithsonian Institution)

Jefferson read his thermometer and made notes of the readings each morning at sunrise or nine o'clock and again at four o'clock or sunset. He recorded wind directions at the same time, and noted snowfall and other climatic events that occurred. During his absences from Monticello, he arranged to have his records of observations maintained by a family member.

Jefferson's greatest annoyance with the British troops after Colonel Tarleton's raid on Monticello seemed to have been not that they had destroyed his livestock and his crops, but that they had broken his barometer, which was almost irreplaceable. In the same period his network of weather observation with the Reverend Madison was interrupted for some time because Madison too had suffered at the enemy's hands. "The British robbed me of my Thermometer and Barometer," he bitterly complained to Jefferson. "We have sent to England and expect a Return by this Spring." To compound Madison's problems, the owner of the thermometer he had borrowed in the interim chose this time to insist on having it returned, making it impossible for Madison to continue his observations.[42]

All forms of meteorological phenomena intrigued Jefferson. He kept abreast of the dramatic meteoric fall over Connecticut in 1807, and in 1818 engaged in a comparative study of climate as deduced from the flowering time of trees conducted in several states. While stationed in France he became interested in the Comte de Buffon's inquiry concerning the comparative moisture of the French and American climates and its probable effect upon animal life. He thereupon had several hygrometers made for him to use for measuring the moisture in the air, one for his own use in France and another to be sent to the United States so that a comparative record could be developed. The instruments did not arrive until after his return to the United States in 1789, however, and the project did not materialize.

Jefferson once envisioned the formation of a huge network of observers in every county in Virginia who would make observations simultaneously. He hoped that by proving its merits, the Virginia network might convince the American Philosophical Society to establish one to maintain records over a period of years in every state. The advent of the American Revolution, however, interfered with this ambitious project, and other priorities thereafter kept Jefferson from returning to it. He never totally discarded the weather recording idea, however, and as late as 1824 he discussed such a plan on a national scale with the object of discovering

A table of thermometrical observations, made at Monticello, from January 1, 1810, *to December* 31, 1816.

	1810.			1811.			1812.			1813.			1814.			1815.			1816.			
	max.	mean	min.	max.	mean	min.	max.	mean	min.	max.	mean	min.	max.	mean	min.	max.	mean	min.	max.	mean	min.	mean of each month.
Jan.	$5\frac{1}{2}$	38	66	20	39	68	$5\frac{1}{2}$	34	53	13	35	59	$16\frac{1}{2}$	36	55	$8\frac{1}{2}$	35	60	16	34	51	36
Feb.	12	43	73				21	40	75	19	38	65	14	42	65	16	36	57	$15\frac{1}{2}$	41	62	40
Mar.	20	41	61	28	44	78	$31\frac{1}{2}$	46	70	28	48	71	$13\frac{1}{2}$	43	73	31	54	80	25	48	75	46
April	42	55	81	36	58	86	31	56	86	40	59	80	35	59	82	41	60	82	30	49	71	$56\frac{1}{2}$
May	43	64	88	46	62	79	39	60	86	46	62	81	47	65	91	37	58	77	43	60	79	$61\frac{1}{2}$
June	53	70	87	58	73	89	58	74	$92\frac{1}{2}$	54	75	93	57	69	87	54	71	88	51	70	86	72
July	60	75	88	60	76	$89\frac{1}{2}$	57	75	91	61	75	$94\frac{1}{2}$	60	74	89	63	77	89	51	71	86	75
Aug.	55	71	90	59	75	85	61	71	87	62	74	92	56	75	88	58	72	84	51	73	90	73
Sep.	50	70	81	50	67	81	47	68	75	54	69	83	52	70	89	45	61	82	54	63	$90\frac{1}{2}$	67
Oct.	32	57	82	35	62	85	39	55	80	32	53	70	37	58	83	$38\frac{1}{2}$	59	76	37	57	73	57
Nov.	27	44	69	32	45	62	18	43	76	20	48	71	23	47	71	20	46	70	24	46	71	$45\frac{1}{2}$
Dec.	14	32	62	20	38	49	13	35	63	18	37	53	18	38	59	12	36	57	23	43	69	37
mean of clear weather.	55			58			55			56			$56\frac{1}{3}$			$55\frac{1}{2}$			$54\frac{1}{2}$			$55\frac{1}{2}$

Jefferson's table of thermometrical observations (from The Virginia Literary Museum, and Journal of Belles Lettres, Arts, Sciences, &c.*)*

meteorological laws, noting that he had completed "a seven years' course of observations intended to characterize the climate of this State."[43]

One of Jefferson's purposes for so assiduously recording information about the weather was to correlate the data with periodic phenomena, such as the breeding and migration of birds, or the appearance, flowering, and fruiting of plants. He attempted to determine whether the large-scale cutting of forests brought about changes in climate, and he investigated the cause and effect of the flood of 1771 and the great snow of the following year, both of which caused considerable damage to his farm. These, the earthquake that shook Virginia two years later, and the killing frost of 1779 were all manifestations of his lifelong preoccupation with weather, its causes, and its effects on men.[44]

Eventually Jefferson's involvement with weather watching resulted in the installation of a weathervane at Monticello. In about 1780 a German officer visiting Monticello wrote: "In his parlour he is creating on the Ceiling a Compass of his own invention by wich he can Know the strength as well as Direction of the Winds." It is much more likely, however, that the vane was on the ceiling of the study or library over the parlor, and not in the ceiling of the parlor itself. When the house was remodeled in 1796, a wind vane was installed in the ceiling of the Northeast Portico, with an index or arrow-shaped hand at the center of a wind rose painted on the ceiling in such a manner that as the vane over the roof moved, the index indicated the wind direction. Jefferson may have been inspired by an aviary or tholus of the villa of the Roman writer Marcus Terentius Varro, which was described in considerable detail in Varro's published work on agriculture, a copy of which was on Jefferson's library shelves.[45]

Most of Jefferson's early observations on weather in Virginia were published later in his *Notes on the State of Virginia*, in which he stated, "Journals of observations on the quantity of rain, and degree of heat, being lengthy, confused, and too minute to produce general and distinct ideas, I have taken five years' observations, to wit, from 1772 to 1777, made in Williamsburg and its neighbourhood, have reduced them to an average for every month in the year, and stated those averages in the following table, adding an analytical view of the winds during the same period."[46]

The East Portico's wind vane and compass (William L. Beiswanger and Ralph Thompson/Thomas Jefferson Foundation, Inc.)

Chapter Three

Astronomy

Measures and rhumbs taken on the spherical surface of the earth, cannot be represented on a plane surface of paper without astronomical corrections, and paradoxical as it may seem, it is nevertheless true, that we cannot know the relative position of two places on the earth, but by interrogating the sun, moon and stars.

—Jefferson to Governor Wilson C. Nicholas, 1819[47]

Among Jefferson's dominant youthful interests was astronomy, which he pursued throughout his lifetime, with an assemblage of fine astronomical instruments that he used in the course of the years. He frequently reported observations he had made to his correspondents, such as of the annular solar eclipse of June 1788, for which he decried his lack of an accurate timepiece. For the purpose of calculating the longitude of his estate, he observed another solar eclipse at Monticello that occurred on September 17, 1788. At various times he ascertained latitudes by means of a box sextant, and he constantly promoted the use of astronomical observations for accurately establishing position in mapping the country and its boundaries.[48]

He had established the latitudes of Poplar Forest and Willis Mountain by his own observations, which were then repeated and verified by his grandson, Thomas Jefferson Randolph. After ascertaining the dip of his Borda circle, Jefferson discovered that he was able to use it on land without having an artificial horizon in the same manner that it was used at sea.[49]

He offered to provide the Reverend Madison with some additional latitudes to use on the map that the latter was preparing, adding,

> *I have a pocket sextant of miraculous accuracy, considering its microscopic graduation. With this I have ascertained the latitude of Poplar Forest, (say New London) by multiplied observations, and lately that of the Willis mountains by observations of my own, repeated by my grandson, whom I am carrying on in his different studies. Any latitudes within the circuit of these three places I could take for you myself, to which my grandson, whose motions will be on a larger scale, would be able to add others.*[50]

On September 17, 1811, accompanied by either James Madison or Payne Todd, among others, Jefferson made observations of the annular eclipse of the sun, recording his data with precision and modesty. He was forced to use an ordinary timepiece instead of a regulator clock, he noted, and he proudly sent copies of his records of the event to Todd and others. He explained that he had perfect observation of the sun's passage over the meridian, and the eclipse had begun so soon thereafter "as to leave little room for error from the time-piece. Her rate of going, however, was ascertained by ten days' subsequent observation and comparison with the sun, and the times, as I now give them to you, are corrected by these." He added, however, that the times of the first and last contacts were not reliable since his clock was too slow, and he had not begun watching soon enough. He reported that nonetheless the last contact was sufficiently well observed but that he had relied on the forming and breaking of the annulus and was certain that there was no error in either. He had not yet calculated the longitude from them, he advised his correspondents, but suggested that Todd could do that as one of his college exercises. He estimated that the latitude of Monticello was 39° 8', and he was correct. He had used his prized "equatorial telescope" for

Jefferson's refracting telescope with double achromatic lens made by Peter and John Dollond (Thomas Jefferson Foundation, Inc.)

Orrery by William Jones, ca. 1790, demonstrating motion of planets in the solar system, similar to one in Jefferson's collection by the same maker. (Thomas Jefferson Foundation, Inc.)

observing the eclipse and noted that one of his companions had observed through a Dollond achromatic telescope, while two other companions attended to the timepieces. Jefferson was pleased by the fact that, as well as he was able to determine, he was the only one in Virginia to have made the observation.[51]

As Jefferson noted, the instrument he used was his Universal Equatorial Telescope made by Jesse Ramsden, the dean of London's mathematical instrument makers. Jefferson purchased it secondhand through David Rittenhouse from John William Gerard De Brahm, a Dutch military engineer and surveyor of the province of Georgia. De Brahm also was appointed to survey the boundary between New Jersey and Delaware, and in 1765 he undertook the survey of the Florida coastline from St. Augustine to the Cape of Florida. Finding himself in financial straits, De Brahm was forced to sell the instrument. Hoping that it would realize a better price than had his other possessions, he left it in Rittenhouse's hands to be sold, where it came to Jefferson's attention. After his return from Washington to Monticello, Jefferson wrote to Rittenhouse, explaining that after having paid all of his bills, he had no funds left for purchasing the instrument before leaving the city; he enclosed a payment including $102.67 for the instrument. In 1821 Jefferson mentioned to an unidentified correspondent that the instrument cost "thirty-five guineas in Ramsden's shop before the Revolution." He was justly proud of his telescope, unquestionably the most sophisticated astronomical instrument, and undoubtedly the only example in the country at that time.[52]

For a long time Jefferson had been planning "to fit up a room to fix my instruments" so that he could determine Monticello's longitude by means of lunar observations, since this method had the advantage "of multiplied repetitions and less laborious calculations." He did in fact manage to provide a place for his instruments in his study upon the sills of the bay window on the southern exposure from which he had a clear view of the sky. He prepared an elaborate "Plan of a Bow window to be made at the Southern angle of my cabinet for observing instruments," which consisted of an extension to be added as an "observatory corner" of the study. The plan featured skylights to be built over both sides of the triangular space of the extension, each of which could be lifted on hinges to open. The plan specified that at the apex of the triangular addition were to be placed three marble slabs at the level of the windowsill, over a solid brickwork pier. The purpose of the pier, which was to extend to basement level, was to protect the instruments from any possibil-

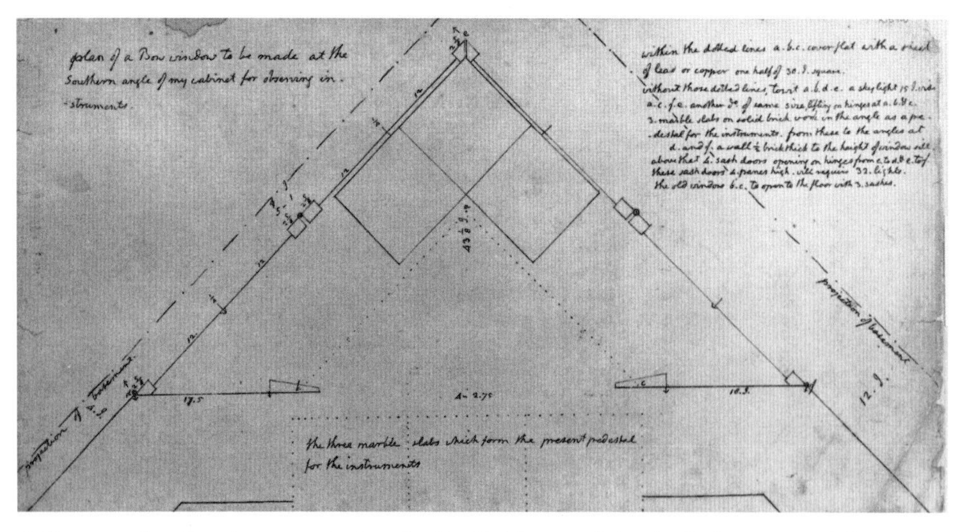

Jefferson's plan for the bow window in his Cabinet at Monticello (courtesy of the Massachusetts Historical Society, N-178, K-154a. Thomas Jefferson Papers, MHS)

ity of vibrations from the floor. From the level of the windowsills, four sash doors each four panes high would enclose the addition on each side, requiring thirty-two lights. They were to be made to open on hinges and the existing windows were to be converted to three sashes and were to extend to the floor. Despite his elaborate planning, Jefferson never managed to find the time to construct the "observatory corner," and instead continued to use the existing platform for his instruments. He had three marble slabs set into the front of the existing window embrasure, supported below by a brick pier extending to the basement for the purpose of eliminating vibrations as he had originally planned.[53]

The statesman had desired for a long time to have a regulator clock of sufficient precision for timing his astronomical observations, and it was after observing an eclipse of the moon in the summer of 1778 that finally he commissioned such a clock to be made. He gave the assignment to David Rittenhouse, whom he had met during the meetings of the Continental Congress in Philadelphia. Rittenhouse was forced to postpone the project, however, being fully occupied with war activities in addition to his responsibilities on the Pennsylvania Committee of Safety. Once again Jefferson put the project aside, and it was not until 1811 that he revived it, ordering—with the assistance of Robert Patterson—a regulator clock from the Philadelphia clockmaker Thomas Voight. Jefferson specified that the clock was to be without ornamentation, housed in a plain mahogany case, "the timekeeping part of which should be perfect .... To be of course without a striking apparatus, as it would be wanted for astronomical purposes only." In addition to using the clock for making observations, Jefferson had in mind to utilize the timepiece also for experimentation with the length of the pendulum as a potential standard of measure.

Initially Voight had quoted a price of $65 for the clock, but by the time it had been completed the price was raised to $115.50. The clockmaker had estimated that a week would be required to regulate the clock before shipment, after which it could be shipped to Monticello. The War of 1812 was at its height at this time, however, and Patterson cautioned that the risks in shipment required insurance, which Jefferson authorized. Then it was realized that the timepiece would have to be shipped on an open boat, exposing it to inclement weather, so

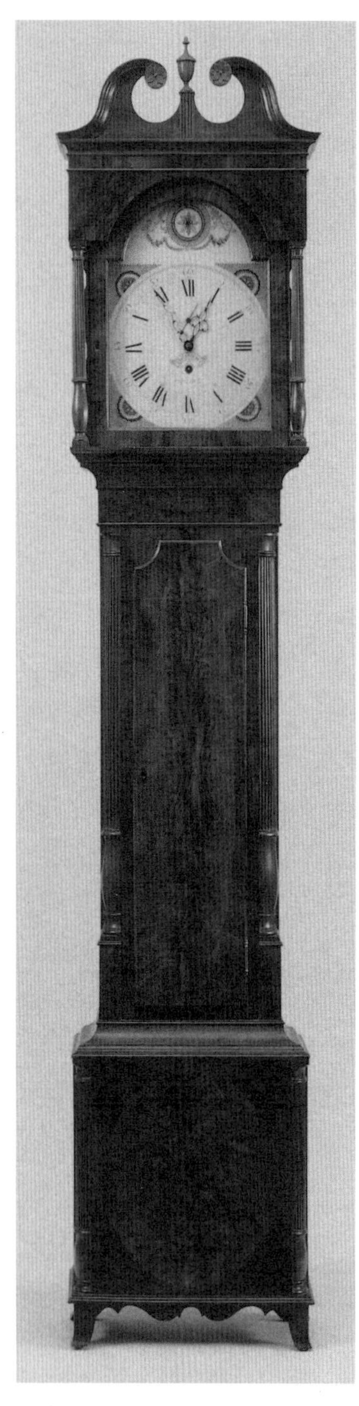

Jefferson's astronomical tall case clock by Thomas Voight, 1812 (Thomas Jefferson Foundation, Inc.)

that special packing was required, which brought about another delay. Then, when the clock was finally ready to be sent to Monticello, it was discovered that the river had become frozen over, and there could be no shipping until it again became free of ice. At just that point when it again became possible to travel, the British fleet blockaded the river. Patterson patiently unpacked the clock in his home again, and at Jefferson's request a rod pendulum was added since he wished to experiment with the rod vibrating seconds. Time passed, during which more modifications were made to the clock, and finally, four years after it had been ordered, the clock arrived at Monticello.

Jefferson placed the timepiece in his study where he kept his telescopes and other instruments. He marked the back board of the clock case for each day of the week so that as the clock weight descended, it marked the current day, serving as a calendar in the same manner as did the Great Clock in the Entrance Hall. Although he used the Voight clock occasionally for astronomical observations, it served him primarily for regulating the times of his rising and retiring for the remainder of his life. He wound it every Sunday morning as he did all the other clocks in the mansion. After his death, Jefferson's daughter, Martha Jefferson Randolph, presented it to Jefferson's physician as a gift. Eventually the clock became part of the collection of the Historical Society of Pennsylvania, but recently it has been returned permanently to Monticello.[54]

Chapter Four

The American Philosophical Society

These societies are always in peace, however their nations may be at war. Like the republic of letters, they form a great fraternity spreading over the whole earth, and their correspondence is never interrupted by any civilized nation.

—Jefferson to John Hollins, February 19, 1809[55]

In his efforts to popularize the sciences, Jefferson found one of the greatest obstacles to be the lack of adequate means to inform the general public despite the efforts of the nation's two scientific societies, the American Philosophical Society and the American Academy of Arts and Sciences. Both attempted to combine research and productivity in the theoretical as well as the practical sciences, although with greater emphasis given to the latter. Jefferson believed that the high cost of import duties inhibited the distribution of European books on these subjects, thus hindering Americans' access to European learning. For this reason he opposed import duties on publications. Jefferson was convinced that the most successful channels for bringing science to the public were the scientific societies, although they did not have the capability to inform society at all levels.

Jefferson was elected to membership in the American Philosophical Society in 1780, and appointed a councilor in 1781 and again in 1783. Chosen as one of the vice presidents in 1791, he was elected president of the society in 1797 to succeed the recently deceased David Rittenhouse. Although he was no longer able to attend meetings after the federal government moved from Philadelphia to Washington, he was reelected to the presidency every year until he resigned from the position on November 23, 1814. At the time of his death, he had been a member of the society for almost forty-six years.[56]

Soon after he was elected vice president, he served on a committee to collect information about the history of the destructive Hessian fly, an insect that was causing more damage to wheat crops throughout the country "than an army of twenty thousand Hessians." The insect, *Phytophaga destructor*, was said to have

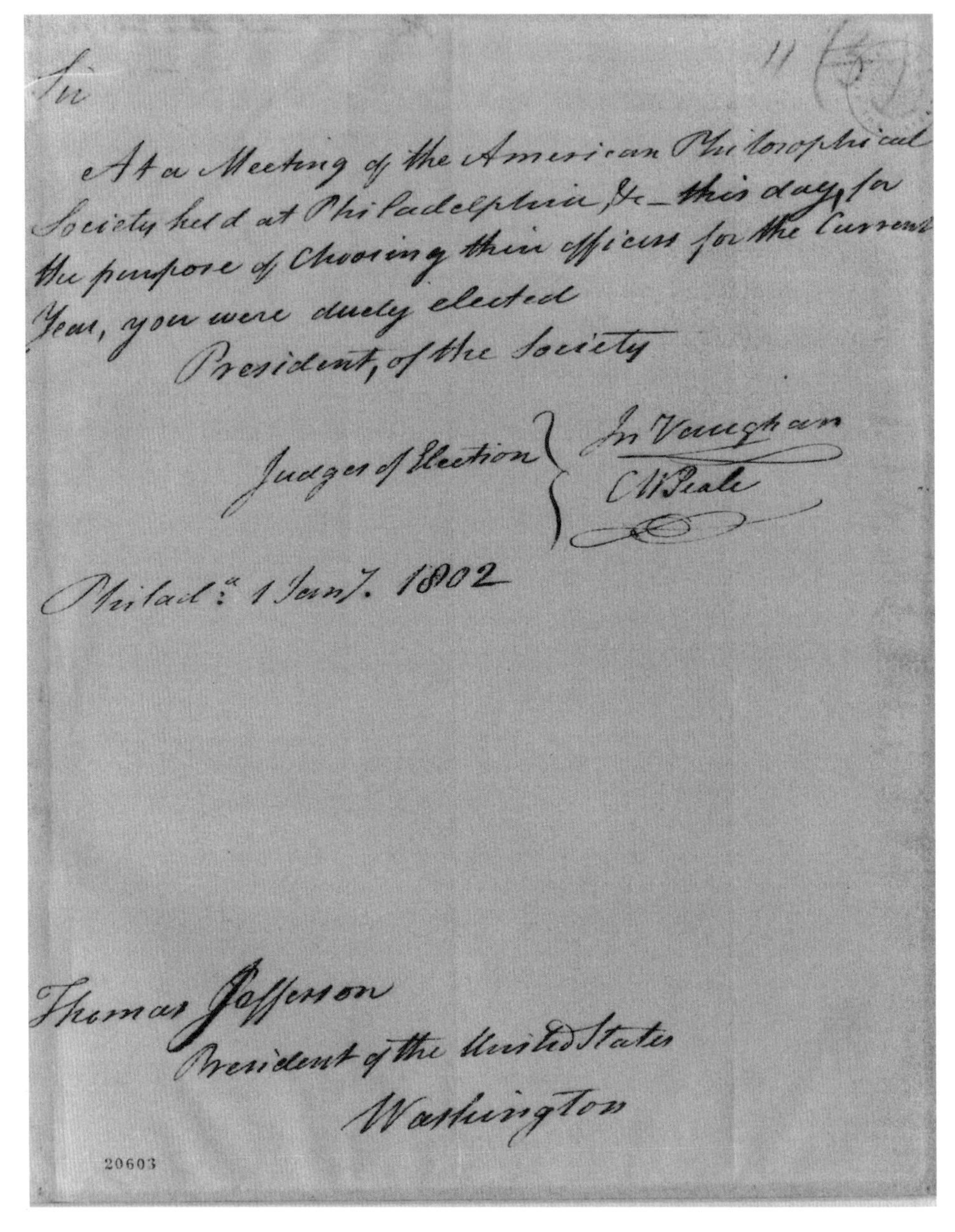

Sir

At a Meeting of the American Philosophical Society held at Philadelphia &c—this day for the purpose of choosing their officers for the Current Year, you were duly elected
President, of the Society

Judges of Election { Jn Vaughan
C W Peale

Philada. 1 Janry. 1802

Thomas Jefferson
President of the United States
Washington

20603

Jefferson's notification that he had been elected president of the American Philosophical Society, January 1802 (courtesy of the Library of Congress)

been introduced into the American continent via hay or straw bedding materials brought by Hessian mercenaries during the American Revolution. To avoid the possibility of infecting their own crops, the British had prohibited the importation of American wheat, a circumstance that seriously affected American farmers. The committee met frequently at Jefferson's home as the members sought a solution. Jefferson compiled a systematic study on the insect and the means for its prevention, based upon interviews with farmers and others. The insect continued to cause damage for the next several decades. It was a study sponsored by President Washington, which for all intents and purposes, was the first entomological survey sponsored by the American government.[57]

Several accounts reporting Jefferson's various scientific endeavors were published in the society's *Transactions*, beginning with meteorological observations that he conducted in conjunction with the Reverend James Madison, which Madison submitted for publication in 1779. Jefferson's other publications were an open letter on methods for calculating the heights of mountains in the Blue Ridge, his description of the moldboard plow, and his account of the paleontological find he named the *Megalonyx*. Jefferson's numerous tangible gifts to the society ranged from ancient Roman bronze coins to various geological, ethnological, and paleontological collections, as well as models of inventions. Jefferson's participation in the society's activities did much to promote the sciences in public service. Notable among Jefferson's contributions to the advancement of American science was his promotion of the study of animal species and of fossil remains.[58]

Chapter Five

Science in Europe

But even in Europe a change has sensibly taken place in the mind of Man. Science had liberated the ideas of those who read and reflect, and the American example had kindled feelings of right in the people. An insurrection has consequently begun, of science, talents, and courage against rank and birth, which have fallen into contempt .... Science is progressive, and talents and enterprise on the alert.

—Jefferson to John Adams, October 28, 1813[59]

Not long after his arrival in Paris, Jefferson was intrigued again and again by new inventions he found that were not known in the United States, which he hastened to confide in letters to Charles Thomson, the Reverend James Madison, and other friends. He was particularly curious about the "cylinder lamp," which, with a small consumption of olive oil, provided a light equal to that of six or seven candles. It was named the "Argand Lamp" after its Swiss inventor, François Pierre Ami Argand. The statesman was equally fascinated by "phosphoretic matches," which he reported were useful for lighting a bedside candle at night, for softening sealing wax, and for replacing flint and steel. Jefferson warned that "great care must be taken ... that none of the phosphorus drops on your hand, because it is inextinguishable and will therefore burn to the bone .... It is said that urine will extinguish it." Thomson informed him, however, that the matches were already known in the United States and were being sold in toy shops in Philadelphia.[60]

In 1785 Jefferson reported having observed a steam engine used for raising water for the fire protection of Paris, and later wrote to Thomson how in London

Imperiale e Reale Accademia Economico-Agraria

dei Georgofili di Firenze

Intenta sempre l'Accademia di ascrivere nel numero de' suoi membri le persone che non solo possono onorarla con il loro nome, ma ancora esserle utili con la comunicazione d'esperienze e di scritti diretti all'avanzamento delle cognizioni teoriche e pratiche riguardanti l'Agricoltura e qualunque altro Ramo d'Economia pubblica e privata, delle Scienze e Arti in quei particolari che con esse abbiano relazione, ha nell'Adunanza ordinaria de' Due Gennajo Milleottocentoventi eletto Voi Eccellenza M.r Thomas Jefferson per uno de' suoi Soci Corrispondenti e perciò Ve ne rimette la presente in prova della vostra ammissione e come un attestato del vostro merito e della nostra stima per Voi.

Data in Firenze nella Residenza dell'Accademia questo dì 15 Gennajo 1820

Vice Presidente

Segretario degli Atti

Certificate of honorary membership presented to Jefferson by the Imperiale e Reale Accademia Economico-Agraria dei Georgofili di Firenze, January 15, 1820 (Thomas Jefferson Foundation, Inc.)

he had learned about the applicability of steam power for operating gristmills. He forwarded to American associates the first news to reach the United States of James Watt's invention of the steam engine, which, he reported, while using no more than one and one half pecks of coal, could perform the equivalent of a horse's full day's work. It was his conclusion, however, that in the United States steam power would lead neither to an industrial revolution nor to a proliferation of factories that would encourage the growth of cities and diminish the importance of agriculture. Steam would become a supplementary source of power, he believed, which would be applicable principally to navigation, milling, small-scale manufacturing, and the performance of everyday chores, with the purpose of liberating men to follow agricultural pursuits. He contemplated various applications of steam to such chores and tried to interest others in them.[61]

"In science, the mass of people is two centuries behind ours," he wrote about the French people to Charles Bellini not long after his arrival in Paris, "their literati half a dozen years before us. Books, really good, acquire just reputation in that time, and so become known to us and communicate to us all their advances in knowledge."[62] During his tour of France and northern Italy in 1787, and of the low countries in the following year, Jefferson found everywhere things that delighted and thrilled him, and he was especially intrigued by technological innovations. He noticed, for example, the "diamond-wise" placement of joists in house construction to provide greater strength to arches between; windows made to admit air while keeping out rain, which he later adapted at Monticello; and in Germany he discovered a bridge over the Rhine River that was being supported by thirty-nine boats. As he traveled, he kept notes of climate, soil, crops, farming implements, and breeds of cattle, all for the purpose of reporting useful information to his countrymen.[63]

His preoccupation with agriculture in all its aspects was a dominant concern. It was his desire to restore agriculture, as he wrote to David Williams in 1803, "to its primary dignity in the eyes of men. It is a science of the very first order. It counts among its handmaids the most respectable sciences, such as Chemistry, Natural Philosophy, Mechanics, Mathematics generally, Natural History, Botany."[64]

Chapter Six

NOTES ON THE STATE OF VIRGINIA

I had always made it a practice whenever an opportunity occurred of obtaining any information of our country, which might be of use to me in any station public or private, to commit it to writing. These memoranda were on loose papers, bundled up without order ... I thought this a good occasion to embody their substance, which I did in the order of Mr. Marbois' queries, so as to answer his wish and to arrange them for my own use.

—JEFFERSON, AUTOBIOGRAPHY[65]

It was during his tenure as governor of the state of Virginia that Jefferson undertook the compilation of the work that was to become his *Notes on the State of Virginia*, one of his major scientific achievements. The French government had expressed a desire to learn more about the country it had assisted during the American Revolution, and the request was transmitted to Chevalier de la Luzerne, the French legate to the United States. The legation secretary, François de Barbé-Marbois, prepared a set of twenty-two questions, to be submitted to each of the states. The questionnaire for Virginia was directed to Jefferson, considered to be the one most knowledgeable about Virginia's geography and flora and fauna as well as its political structure. Even before he left the governorship, Jefferson began studying the questionnaire and gathering his materials, and following his retirement, he turned his attention to the project in earnest.

Jefferson structured his response on the format of the questionnaire, and his compilation included a description of Virginia's geography, its topographical features, waterways, climate, natural resources, minerals, boundaries, and flora and fauna. He described the state's population, commercial development, military forces, and social and political life. He vigorously refuted assumptions made and published by the distinguished French naturalist, Georges Louis Leclerc, Comte de Buffon, concerning the American Indians and American flora and fauna. He disputed Buffon's conclusion that animals common to both the Old World and the New were of smaller scale in the latter. The French naturalist claimed that animals domesticated in America had degenerated and that in general fewer species were to be found in the New World than in the Old. In refutation of Buffon's opinions, Jefferson later sought to prove the error of Buffon's contentions by sending him the skeleton, hide, and horns of a moose as well as examples of other animals.[66]

Ever jealous of the prestige of the new republic of which he had been an architect, Jefferson marshaled facts and arguments to refute Buffon's contentions. He also argued against the proposal made by Louis Daubenton, one of Buffon's associates, that the remains of the "mammoth" found on the North American continent were of two different species and that they belonged to either the rhinoceros or the elephant. Jefferson discarded both classifications and proved conclusively that they were the remains of the mastodon and that it was a different species from the Siberian mammoth.[67]

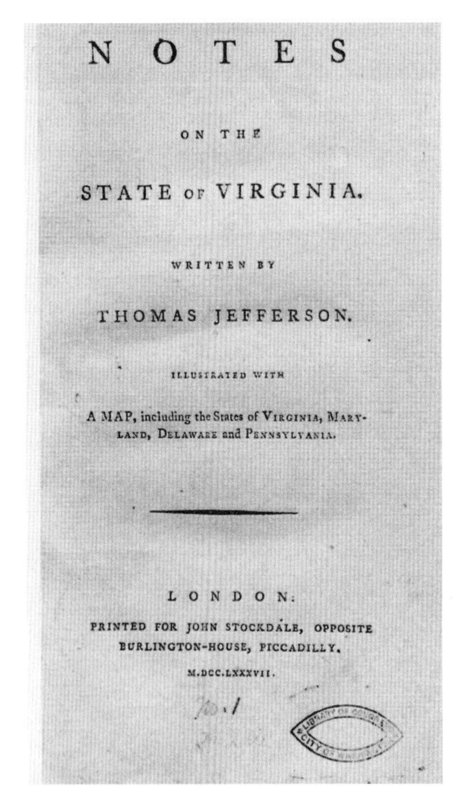
NOTES
ON THE
STATE OF VIRGINIA.
WRITTEN BY
THOMAS JEFFERSON.
ILLUSTRATED WITH
A MAP, including the States of VIRGINIA, MARYLAND, DELAWARE and PENNSYLVANIA.
LONDON:
PRINTED FOR JOHN STOCKDALE, OPPOSITE BURLINGTON-HOUSE, PICCADILLY.
M.DCC.LXXXVII.

Title page of the first English edition of Notes on the State of Virginia *(courtesy of the Library of Congress)*

He was in error, however, in supposing that the mammoth or mastodon still survived somewhere in the Mississippi valley or beyond, contending that "such is the economy of nature that no instance can be produced of her having permitted one race of her animals to become extinct." He was careful, however, to not assume the other extreme and claim that the animal life of America was superior to that of Europe.[68]

He particularly took issue with Buffon's detrimental estimate of the American Indians, and by contrast emphasized their virtues. The

Indians, wrote Jefferson, were "led to duty and to enterprise by personal influence and persuasion. Hence eloquence in council, bravery and address in war, become the foundations of all consequence with them." From Buffon Jefferson learned the important lesson that hazards always lay in making categorical assertions in scientific matters. He believed that he had convinced Buffon of his errors, and the French naturalist promised to make some corrections in his next volume. He died before he was able to do so, however.[69]

The first evidence of Jefferson's interest in and familiarity with the natural sciences became evident with the compilation of his *Notes*. Despite his dedication to the study and practice of law and later to his career in politics, he managed to amass nearly all the knowledge of geology and zoology available in his time. He was familiar with the published works of the living European scientists, and knew more about the animal life of North America than did any of his contemporaries. Although modern science has rejected most of the theories that Jefferson projected in his *Notes*, he was nonetheless in advance of the best specialists of the time in his conclusions on paleontology. No complete theory of the origin of petrifactions occurring in the earth's strata had as yet received general acceptance. Linnaeus had classified fossils among minerals while others considered them freaks of nature.

Although he was not an experienced cartographer, Jefferson had produced one map for public use early in his career. He had experience as a surveyor, and as part of his law practice, he had drawn up land plats used in litigations. Now in order to create a new, updated map, he planned to combine the several existing maps of Virginia and supplement them with new information he expected to obtain from various sources. He laid out his map with the prime meridian at zero degrees at Philadelphia, in accordance with common usage. He included an area of about one degree west of Philadelphia and from thirty-six degrees to forty-two degrees north latitude. Regions added were those of North Carolina north of Albemarle Sound, all of the states of Virginia, Maryland, Delaware, and Pennsylvania, the western regions of New York and New Jersey, and the territory westward just beyond where the Ohio and Great Kanawha Rivers met.[70]

Jefferson utilized recent cartographic works available to him, which he knew with certainty had been produced from actual surveys and consequently

NOTES on the ſtate of VIRGINIA;
written in the year 1781, ſomewhat corrected and enlarged in the winter of 1782, for the uſe of a Foreigner of diſtinction, in anſwer to certain queries propoſed by him reſpecting

MDCCLXXXII.

1431

Table of contents of Jefferson's Notes *(courtesy of the Library of Congress)*

were the most reliable. Among them was an available version of the William Scull map of 1770 for the section on Pennsylvania, and for Virginia he utilized the map that had been produced in 1750 by his father and Joshua Fry. He revised it with more recent additions from the 1778 survey made by Thomas Hutchins. Finally, he added information about the western regions he had collected from explorers and land developers, and included geographical phenomena not previously recorded cartographically, such as the Natural Bridge, caves, and the Indian mound he had excavated. He also divided the western lands into five new states, of which he named three, and indicated what he believed to be the correct western boundary of Virginia. For platting latitudes, he used observations he had made at Monticello and those made at Williamsburg by the Reverend Madison. He obtained the latitudes of the western limits of Pennsylvania from David Rittenhouse. Using a scale of one inch to twenty miles, he worked on the map for more than a year, completing it in the spring of 1786.[71]

The *Notes* provided ample evidence of Jefferson's vast erudition. He cited eighteen foreign authorities, quoting and translating from four foreign languages. The work, intended for distribution only to friends and colleagues, was privately printed in 1785. Subsequently reprinted in England, France, and the United States in numerous editions, it was the first comprehensive study to be made of any part of the United States and is now considered to be the most important scientific work published in America in the eighteenth century.[72]

Chapter Seven

ETHNOLOGY

In the early part of my life, I was very familiar [with the Indians] and acquired impressions of attachment and commiseration for them which have never been obliterated.

—JEFFERSON TO JOHN ADAMS, JUNE 11, 1812[73]

While he was growing up as a boy at the edge of the Virginia frontier, Jefferson became personally acquainted with Indians whom his father had befriended when they stopped on their way to Williamsburg. Later, during the years that he was a student at the College of William and Mary, he visited the camps of Indians who came to the Virginia capital. While he was governor of Virginia and then president of the United States, Jefferson frequently received visits from the Indians, during which he sought to encourage them to agricultural pursuits.[74]

With his interest in the Indians, Jefferson had long been curious about the puzzle of the Indian mounds, or "barrows," as he called them. He was moved to action, however, by a query regarding "Indian monuments" in the questionnaire that had prompted Jefferson to write *Notes on the State of Virginia*. Sometime around 1782, during the period when he was compiling the *Notes*, Jefferson undertook an excavation of an Indian mound in his own neighborhood. The site was approximately two miles above the principal fork of the Rivanna River, opposite several hills where former settlements of the Monocan Indians were known to have existed. It was situated on the site of the ancient Monocan village of Monasukapanough, along the low lands beside the Rivanna River near his home. It was not an endeavor for the purpose of collecting artifacts, but a serious effort

Earthenware figurine, heads, and ball rattle believed to be from Jefferson's collection, dating to the Mississippian period, AD 900–1550 (Thomas Jefferson Foundation, Inc.)

to resolve conjectures concerning the structure and purpose of these aboriginal burial places. Jefferson proceeded methodically with his excavation, recording the evidence encountered as well as the stratigraphy, in a remarkably professional and even modern manner.

In his account he explained that he was unaware of any existing Indian monuments but made an exception of the "barrows," or tumulus graves now called "Indian mounds," of which many were to be found in the country. Built of earth or loose stones, they were constructed in various sizes as repositories for the bodies of the dead. While nothing was known as fact, some conjectured that they were the graves of fallen braves buried on the site of combat. Others suggested that they were community graves in which the bones of all the Indian dead within a specified period had been collected from numerous individual graves. Yet others, as Jefferson explained, believed that they were "general sepulchers for towns."

The mound had a "spheroidal" shape, approximately forty feet in diameter at the base, and Jefferson concluded that it may have been as much as twelve feet in height. Unfortunately, the soil had been plowed during the previous decade, which he estimated had reduced the level as much as seven and one half feet. There was evidence that trees with trunks as large as twelve inches in diameter had existed on the site before plowing. All appearances pointed to the fact that the mound had not been an occupied site but an accumulation of bones from ancient burials, the first of which had been deposited directly on the ground and to which others had subsequently been added in layers, with a few stones placed over each burial.

Jefferson next related the mound to its natural surroundings and surviving evidence of human occupation of the area, then observed and correctly interpreted the stratigraphical stages of his excavation. He appeared to have had no hesitation about desecrating a human burial, presumably because he felt that the knowledge to be gained would justify it. He undertook the excavation as a scientific inquiry. By applying his innate sense of order and detail, he anticipated modern archaeology's basis and methods by almost a full century. Though working during a period when there was no established procedure for excavating ancient sites, he followed the best available model for the project. It was not until the nineteenth century that the science of archaeology began to evolve slowly within the ambience of the academic world.

Buffalo hide shield representative of Native American objects at Monticello (Thomas Jefferson Foundation, Inc.)

The importance of Jefferson's experience and his report of it cannot be overstressed, for he correctly used stratigraphy to make inferences about the past—a century before the principle became a basic part of the methodology of all archaeology, regardless of where it is undertaken. The principle is the method for providing a calendar for establishing the age of remains. As the criterion of scientific excavation, it is the principle in which every student of the science is to be trained. As C. W. Ceram, the noted author in archaeology, commented, Jefferson "not only

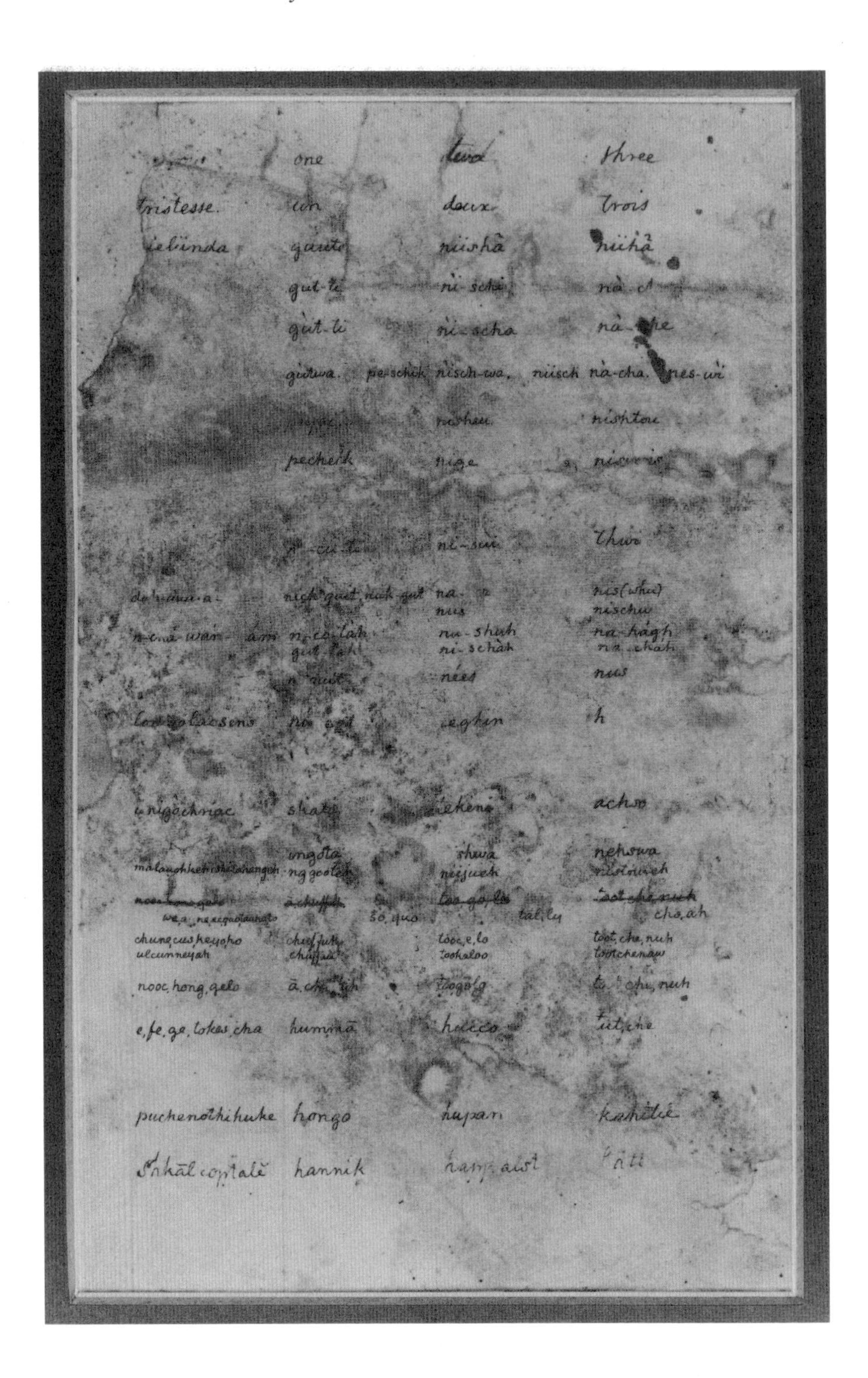
one two three
tristesse. un deux trois
puchenothihuke hongo hupan kahtilee
hannik

Page one of Jefferson's Indian vocabularies (courtesy of the American Philosophical Society)

indicated the basic features of the stratigraphic method but also virtually named it, although a hundred years were to pass before the term became established in archaeological jargon."[75]

Jefferson eventually reached the conclusion that the best record of Indian origins rested in their languages, and he contemplated a major compilation of Indian vocabularies. He began collecting Indian vocabularies in about 1779 or 1780, according to his own statements, probably while engaged in compiling his *Notes*. "At a very early period of my life," he explained to Levett Harris, "contemplating the history of the aboriginal inhabitants of America, I was led to believe that if there had ever been a relation between them and the men of color in Asia, traces of it would be found in their several languages."[76]

During his tenure as president of the American Philosophical Society and under his leadership, a Standing Committee on Antiquities was appointed "to inquire into the Customs, Manner, Languages and Character of the Indian Nations." As a young man Jefferson designed a vocabulary in which he included frequently used words along with names of commonplace objects that were to be found everywhere and which consequently would have a name in every language. Included were 250 common words found in a number of Indian languages, and in an attempt to discover a relationship between them, Jefferson juxtaposed European equivalents to the Indian words. In the course of compiling his vocabularies, Jefferson interviewed Indian traders and guides whom he met, and by means of correspondence he consulted Indian agents afield. He also urged explorers going into Western lands, notably Lewis and Clark, to collect and record whatever they could about Indian languages. In his *Notes* he proposed that the Indian languages be arranged "under the radical ones to which they may be palpably traced, and doing the same by those of the red men of Asia, there will be found probably twenty in America, for one in Asia, of those radical languages, so-called because if they were ever the same, they have lost all resemblance to one another." Using this compilation as a basis, he produced a standard form for Indian languages which was then printed and distributed widely by the society to anyone having contact with aborigines and the willingness to complete the forms.[77]

In preparation for his departure from Washington to Monticello following his retirement from the presidency, Jefferson collected all the copies of the vocabularies that had been completed, and packed them together carefully in a trunk with other papers. The trunk was shipped by water with thirty other packages of his effects. When the vessel was ascending the Jamestown River, a thief seeing the trunk assumed that because of its weight that particular trunk must have contained valuables. When he opened it, however, he was disappointed to find only papers and a few miscellaneous objects of no market value and threw the trunk's contents into the river. Some pages of the vocabularies floated ashore and were recovered from the mud. They were so greatly damaged that for the most part they were useless, and Jefferson managed to save only a few.[78]

Replication of an otter bag used for carrying ceremonial tobacco in the time of Lewis and Clark (Thomas Jefferson Foundation, Inc.)

Jefferson collected Indian objects for many years, particularly during the period that he was involved with the preparation of his *Notes*. It was not until much later, after he remodeled Monticello, that he mounted the collection in the new Entrance Hall, together with fossils, geological specimens, and other natural history materials. The impetus for the creation of this display was Jefferson's receipt in August 1805 of a large shipment of Native American objects and natural history specimens from Lewis and Clark. Included in the collection were Indian bows, arrows, quivers, lances, pipes, wampum, beaded or quilled moccasins, dresses, and cooking utensils of the Mandan and other Indian nations of the Missouri. The room also featured two painted buffalo hides, one showing a map of the Missouri River and its tributary streams, and the other a representation of a battle between the Mandan and their allies against the Sioux and Arikaro.

Chapter Eight

Health and Medicine

While surgery is seated in the temple of the exact sciences, medicine has scarcely entered its threshold. Her theories have passed in such rapid succession as to prove the insufficiency of all, and their fatal errors are recorded in the necrology of man. For some forms of disease, well known and well defined, she has found substances which will restore order to the human system, and it is to be hoped that observation and experience will add to their number.

—Jefferson to Dr. John Crawford, January 2, 1812[79]

One of the scourges of the eighteenth century was smallpox, of which epidemics occurred periodically in both the Old World and the New. It was estimated that the case fatality rate ranged from 20 to 80 percent since 1650. While epidemics were ravaging white communities, others occurred also among Indian populations in Canada and the American West, often devastating as much as half the population of a tribe. The recurring epidemics followed in the wake of the African slave trade, damaging black populations as much as American Indian tribes. Although inoculation had been introduced in England and in the American colonies at almost the same time, it was forbidden in one colony after another in the belief that it spread the disease instead of preventing it. Opposition to inoculation by most physicians resulted in a battle between the clergy and the medical community, often leading to mob violence.[80]

Reports of the work being done in England by Dr. Edward Jenner came to the attention of Dr. Benjamin Waterhouse at Harvard, who became interested in the subject of vaccination as prevention of the disease. He appealed to his friend

President John Adams for assistance in making the procedure known throughout the country, but Adams took no action in the matter.[81]

Then Jenner applied to Vice President Jefferson, who promptly supported Waterhouse in his efforts to develop his program of vaccination and personally provided a means for distributing the vaccine to physicians in Philadelphia, Virginia, Washington, and elsewhere. When Waterhouse sent him some vaccine, Jefferson turned it over to Dr. Edward Gantt, chaplain of the Congress, to distribute.[82]

Jefferson also forwarded some vaccine with carefully detailed instructions for its use to his family physician, Doctor William Wardlaw, in Albemarle County. Wardlaw inoculated six members of Jefferson's family, but his medical

Small portable medicine chest, ca. 1800 (courtesy of Mrs. Prentice Cooper)

practice was too demanding for him to continue. Thereupon Jefferson and his two sons-in-law personally continued the procedure in Virginia. Jefferson subsequently reported having personally inoculated from seventy to eighty members of his own family and household, including slaves, during the summer, and his sons-in-law had inoculated approximately as many of their own families and neighbors. Between them they had achieved a combined total, he reported, of approximately two hundred who had been vaccinated, of which only one case was attended by fever. Jefferson also worked out an improved system for the transportation of the vaccine, which Waterhouse adopted.[83]

Anxious to expand the sphere of inoculation, Jefferson forwarded some of the vaccine that he had obtained from Waterhouse to the Philadelphia merchant John Vaughan to pass on to Dr. John Redman Coxe, already known as an advocate of inoculation. Coxe became the first to introduce inoculation in Philadelphia, and in 1802 he published a book detailing his procedures and experiences with smallpox.[84]

Jefferson was particularly mindful of the epidemics decimating Indian populations, and in December 1804, when an Indian delegation led by the Miami chieftain Little Turtle arrived in Washington to meet with the Congress, Jefferson invited the Indian leader to the White House, stating that he had a message to convey to him that was of importance to the entire Indian nation. As later recounted by Waterhouse, during the meeting with Little Turtle, Jefferson informed him "that the Great Spirit had lately made a precious donation to the enlightened white men," and he then went on to explain the history of vaccination, which was a gift from heaven that would preserve them from smallpox. Little Turtle, who had been listening with close attention, then requested to be the first of his tribe to receive the benefits of inoculation. Dr. Gantt proceeded to inoculate the chieftain and then nine or ten warriors in his entourage. Before they departed, Jefferson provided them with instructions and an amount of the vaccine to take with them. Later, during the preparations for the Lewis and Clark expedition, Jefferson provided the exploring party with a quantity of vaccine material and instructions that they were to vaccinate as many as possible of the Indians they would encounter in the course of their expedition.[85]

Chapter Nine

Paleontology

Wherever these grinders are found, there also we find the tusks and skeleton .... It will not be said that the hippopotamus and elephant came always to the same spot, the former to deposit his grinders, and the latter his tusks and skeleton .... We must agree then that these remains belong to each other, that they are of one and the same animal.

—Notes on the State of Virginia, 1787[86]

Although Jefferson has often been called the father of American vertebrate paleontology, the attribution has been disputed from time to time because he was not the first to collect and scientifically study paleontological remains in the American colonies and new republic. Nonetheless, he personally was largely responsible for popularizing the subject and for preserving many specimens that otherwise might have been lost. Like most of the important men of science in his time, his endeavors were necessarily limited by his occupation. He was to be numbered among those about whom Benjamin Smith Barton wrote in 1804, "it is ever to be regretted that the principal cultivators of natural science in the United States, are professional characters, who cannot without essentially injuring their best interests, devote to these subjects that sedulous attention which they demand ... in *some* respects, they are certainly better qualified to undertake and to perform, the task than the naturalists of Europe."

It was Jefferson who was in part responsible not only for popularizing the science in the United States, but also for making it a respectable pursuit, and for assembling many of the materials that were responsible for its advancement. Not only did he search for and collect fossil specimens over a period of several decades, but he also supported the search for other specimens by others.[87] Jefferson began his collection of fossil remains at the time that he was writing the *Notes*. He acquired specimens from several sources, and he requested explorers such as George Rogers Clark and James Steptoe to obtain fossil remains for him.[88]

During his residency in Paris as minister to France, Jefferson utilized every opportunity to learn about the state of the sciences by studying the natural history collections of the Cabinet du Roi and other collections assembled in Paris. He became acquainted with Europe's foremost scientist, Comte de Buffon, and his associate Louis Daubenton, along with some of the younger French scientists, including Bernard Lacépède, André Thouin, and Faujas de Saint-Fond. He corresponded with the younger men after his return to the United States, and as a consequence of these associations, he later presented an extensive collection of American fossils to the Musée National d'Histoire Naturelle in Paris.[89]

After Jefferson's return to the United States and while serving as secretary of state, he continued his search for and acquisition of the fossil remains that were being found in various parts of the country. In 1797, while excavating the floor of a cave on the premises of Frederic Cromer in Greenbrier County, Virginia (now West Virginia), laborers recovered a number of bones from several feet below the

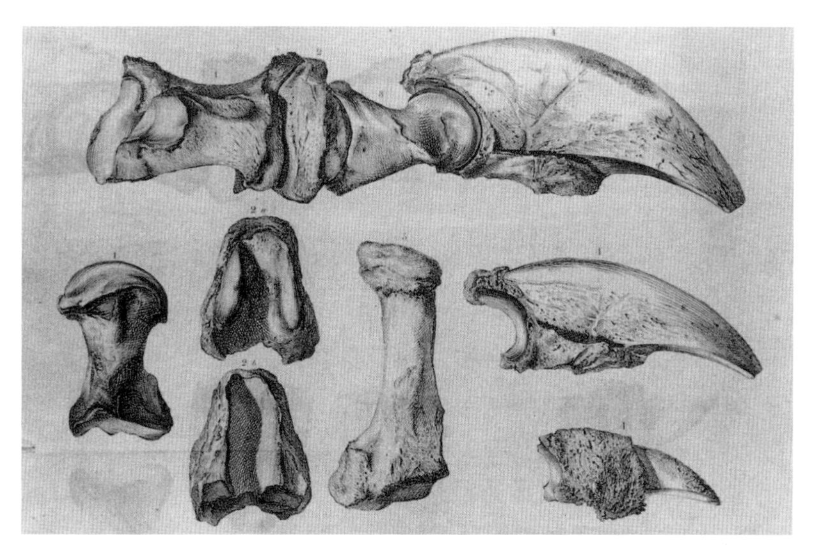

Claws and related bones of the Megalonyx jeffersonii *presented by Jefferson to the American Philosophical Society (courtesy of the Academy of Natural Sciences of Philadelphia, Ewell Sale Stewart Library)*

surface. Upon learning of the discovery, Jefferson obtained the remains through a friend, expecting the bones to be those of a mammoth. To his surprise, they proved to be of the foot of an animal not previously known. Because of the size and characteristics of the claws, Jefferson concluded that they belonged to a carnivore of the family of a lion or a tiger, and he named it "the Great Claw," or *Megalonyx*. He prepared a paper to be read before the American Philosophical Society in which he described the bones, and on his journey from Monticello to Philadelphia to assume the vice presidency of the United States, he brought the bones with him and presented them to the society.

A short time later Jefferson read an article in an English periodical reporting the recovery in Paraguay of the skeleton of an animal called a *Megatherium*, a species of extinct ground sloth. He immediately recognized the possibility that his *Megalonyx* might be similar to the *Megatherium* and then quickly revised his paper before it was published in 1799 in the society's *Transactions*. Later Dr. Caspar Wistar identified the remains as the bones of an extinct species of a large ground sloth, which he named *Megalonyx jeffersonii*.[90]

An obvious center for natural history studies and public information was the museum, and it is not surprising that Jefferson provided eager support to the first successful American museum enterprise, founded by Charles Willson Peale in Philadelphia in June 1784. Jefferson enjoyed a long and fruitful friendship with Peale and provided his museum with numerous specimens over time. Peale's museum was unique as a consequence of its success, for in the eighteenth century museums intended for the edification and instruction of the public were generally short-lived. Men of science for the most part were reluctant to help promote the unusual and were distrustful in general of museums. The public was not yet oriented to museums and consequently did not adequately support them. The favorable response enjoyed by Peale's enterprise was due largely to his great talent as an entrepreneur and his unflagging efforts to provide novel and attention-getting dimensions to his displays. Nevertheless, he failed to engender sufficient support in scientific circles, and he could not convince the federal government to convert the museum into a national institution.[91]

In 1807 Jefferson commissioned William Clark to return to Big Bone

Lick to collect more specimens for him, which he intended to present to the American Philosophical Society for the purpose of forming a major study collection. In January of the following year, after working on the site with laborers for several weeks, Clark shipped three crates containing some three hundred fossil remains to the President's House. Jefferson unpacked and arranged the bones in his "study," one of the unfinished rooms that had become the repository for his gardening and carpentry tools and collections of artifacts. Now the East Room of the White House, during Jefferson's occupancy it was described by one of his visitors: "Around the walls were maps, globes, charts, books, &c. In the window recesses were stands for the flowers and plants which it was his delight to attend and among his roses and geraniums was suspended the cage of his favorite mocking-bird."[92]

Jefferson invited the Philadelphia anatomist Caspar Wistar to Washington to study the collection for the purpose of making a selection to be presented to the American Philosophical Society. Working together in his study, Jefferson and Wistar divided the fossils into three groups. The first, which was the most complete, was set aside for the society. A second selection was earmarked for Jefferson's personal collection at Monticello, and the remainder, which contained many duplicates, Jefferson shipped to the Musée National d'Histoire Naturelle in Paris.[93]

The collection that Jefferson sent to Philadelphia was maintained in the society's Cabinet, and in February 1809 Wistar presented his observations on the fossils before the society. Although the paper was scheduled for publication, the manuscript was lost by the committee of referees. Wistar rewrote his paper in part, and the beginning of his description of the collection was published in the *Transactions* of 1818. He died before it appeared, however, and his other observations on the fossils were never published.[94]

In 1840 the society took the first step towards the disposal of its museum collections. The fossil organic remains in its holdings were transferred to the Academy of Natural Sciences of Philadelphia, with the condition that appropriate means be taken for their preservation. Subsequent deposits were made to the academy by the society in 1852 and again in 1860, and eventually all of the natural history collections that the society had received from Jefferson became part of the academy's collections. Notable among the Jefferson fossils are the eleven bones of the foot of the *Megalonyx jeffersonii*, a tooth of *Symbos cavifrons*, the jawbone and teeth of *Mammut americanum*, and a type specimen of *Tetracaulodon jeffersonii*, as well as several dozen other specimens.

Although Jefferson continued his scientific interests to his final days, in his last years he was completely preoccupied with planning and building the University of Virginia, and he was no longer able to continue collecting due not only to his advanced age, but also lack of funds. In acknowledgment of his contributions, Jefferson was elected a corresponding member of the academy in 1818. When informed of his election, Jefferson expressed his appreciation of the honor, writing, "At an earlier period of life, I might have endeavored to deserve it in fact, but now can only do it by good wishes for its success, & by assurance that I should be gratified by any occasion of being useful to it."[95]

Jawbone of American mastodon (Mammut americanum), *excavated in 1807 and among many fossils and natural artifacts in Jefferson's collection (courtesy of the University of Virginia)*

Chapter Ten

Exploration and Discovery

A great deal is yet wanting to ascertain the true geography of our country; more indeed as to its longitudes than latitudes. Towards this we have done too little for ourselves and depended too long on the ancient and inaccurate observations of other nations.

—Jefferson to Andrew Ellicott, June 24, 1812[96]

In 1803, even before negotiations for the Louisiana Purchase had been completed, Jefferson fulfilled at last his long-held dream of western exploration. President Jefferson obtained the approval of Congress to send an expedition under the leadership of Meriwether Lewis, Jefferson's private secretary from 1801-3. In instructions to Lewis dated June 20, 1803, Jefferson writes, "The object of your mission is to explore the Missouri river, & such principal streams of it, as, by it's course & communication with the waters of the Pacific Ocean, may offer the most direct & practicable water communication across the continent, for the purposes of commerce."[97]

The resulting Lewis and Clark Expedition, described as "the most consequential and romantic peace time achievement in American history," had its genesis in Jefferson's mind fully two decades before the exploring party departed westward from Pittsburgh on August 31, 1803. The need to determine the character and expanse of the western regions of the continent lingered in his mind, and during the intervening years he encouraged three unsuccessful attempts to explore them. Now as president, he was finally able to realize his dream. Not only did the venture achieve all that he had hoped, but also it was the first and one of the most important applications of scientific practices and instrumentation attempted by the young republic. The third president was eminently suited and equipped to plan and direct such a project, for he was better informed on national geography than anyone else in the country. He had spent many years collecting and studying all that had been written and published on the subject. He had opportunities to meet Indians, traders, and others who had traveled to the West, and he recorded what he learned from them. He was knowledgeable about scientific practices and instruments, and was experienced in surveying, mapping, and making astronomical observations, all of which would be required to record the regions to be explored. Furthermore, as president of the American Philosophical Society, he was in a position to call upon the nation's most eminent men of science for detailed advice on all subjects involved in the preparation of the proposed expedition.[98]

To lead the expedition, Jefferson selected his own secretary, Meriwether Lewis, who was self-taught in the natural sciences and whose army career provided experience in handling men. Jefferson had definite opinions about not only the scientific data to be collected but also the means to be used. For advice on the scientific observations to be made, Lewis was sent to be instructed by Professor Robert Patterson, an authority on scientific principles and instruments at the University of Philadelphia, and for instruction in surveying, Lewis went to the nation's foremost surveyor, Andrew Ellicott at the Pennsylvania Land Commission in Lancaster. Jefferson was particularly insistent that "all observations are to be taken with great pains & accuracy," and that they were to be recorded distinctly and in such a manner that they would be comprehensible to others who might follow. Upon the return of the exploring party, these records were to be submitted to the War Office for final calculations to be made by several qualified individuals. To ensure against loss of records, Jefferson directed the explorers to use their leisure time to make several copies of their notes, giving the copies to their most trusted men for safekeeping. "A further guard," he wrote, "would be that one of these copies be on the paper of birch, as less liable to injury from damp than common paper."[99]

Jefferson also required the explorers to compile records of climatological data similar to the records he maintained daily. They returned with a remarkable corpus of information on the flora, fauna, ethnology, and geography of the regions they had traversed. As Jefferson wrote to William Dunbar in Mississippi in 1805, while the expedition was still in progress, "The work we are now doing is, I trust, done for posterity, in such a way that they need not repeat it .... We shall delineate with correctness the great arteries of this great country. Those who come after us will extend the ramifications as they become acquainted with them, and fill up the canvas we begin."[100]

Elk antlers acquired by Lewis and Clark (courtesy of the University of Virginia)

The successful completion of the Lewis and Clark expedition made possible government sponsorship of other similar ventures during Jefferson's administration. In the years 1805 and 1806 an exploring party under Zebulon M. Pike was sent to explore the sources of the Mississippi, Arkansas, and Red Rivers, as well as the western part of the Louisiana Territory as far as Colorado Springs, the home of what we now know as Pike's Peak. Other geographical and geological expeditions followed in fairly rapid succession, leading first to the creation of an Army unit in 1812 that would eventually become the Corps of Topographical Engineers. Ultimately, the United States Geological Survey was formed in 1879.[101]

When Jefferson began sending out expeditions to explore and map North America, he became aware of the need for a single prime meridian to consolidate the information the explorers would collect. The matter of an American prime meridian preoccupied Jefferson for a period of time. It was required not only for establishing longitude on land, but also to serve the American maritime trade. A national prime meridian, not dependent upon any other country, would be a major step in making the new republic completely self-reliant. He was determined to avoid an unnecessary dependence upon Great Britain, and furthermore, he reasoned, a prime meridian at Greenwich, England, was too distant, given the expanse of the American continent. Yet another reason, and one of more immediate personal concern to him, was that the city of Washington, the nation's capital, lacked any facility for time regulation. "There is no such thing as a meridian or other means of keeping our clocks & watches right at this place," he grumbled to Robert Patterson.[102]

The "Jefferson Stone" in front of the Washington National Monument (photo by Eric Long, Smithsonian Institution)

When the surveyor Isaac Briggs, a friend whom Jefferson recently had appointed Surveyor General of the Mississippi Territory, visited Washington in early 1803, Jefferson took advantage of his presence and assigned him the task of establishing the meridian. Briggs determined that "the true meridian line" was to pass in a north-south direction through the center of the President's House, the line to be perpetuated to the point of intersection with a line due west that had been drawn from the center of the Capitol building. The point of intersection was to be identified by means of a permanent monument in an area south of the President's House that provided a magnificent view of the Potomac River.[103]

The meridian marker became identified as the "Jefferson Stone" and was supplemented by two other markers. The nearby "Capitol Stone" indicated the intersection of the north-south line through the President's House with the east-west line through the southern end of the Capitol building. The "Meridian Stone" had been erected on the meridian line north of the President's House on Peter's Hill, now Meridian Park. Upon completion of the work, Briggs prepared a written detailed account of the survey, which was submitted to President Jefferson and subsequently filed. No further action was taken by the president to establish a national prime meridian, presumably due to precedence of other priorities in government.

As the years passed, the Jefferson Stone proved to be a convenient guy post for tying up barges and vessels that traveled the Canal; it was also well situated for surveyors laying out land parcels in the area. In 1848 construction of the Washington National Monument was begun nearby, and the Jefferson Stone was useful in determining potential settlement of the monument as well as any changes in alignment in the construction of the Washington Aqueduct, then also in progress. When work on the monument ceased due to lack of financial resources in 1856, the monument grounds reverted to a neglected and unsightly area.[104]

At the conclusion of the Civil War, the Corps of Engineers were assigned the task of cleaning up the city. In the course of their work, and being uninformed of the purpose of the markers, they removed and discarded both the Jefferson Stone and the Meridian Stone. When work on the National Monument was resumed after 1876, the line was rerun using what was erroneously presumed to be the Jefferson Stone, revealing an alarming degree of unequal settlement of the structure. It was soon discovered that the benchmark presently being used was the Capitol Stone, and it was no longer possible to check the alignment of the monument or the Washington Aqueduct. After much confusion and many difficulties, the Jefferson Stone's original position was found and a new granite marker was installed in 1889.[105]

For more than a century the restored Jefferson Stone has remained staunchly in place a short distance northwest of the Washington National Monument. Its lonely and forlorn appearance continues to mystify residents and tourists alike, and few are aware that it is the only surviving relic of Jefferson's plan to establish a national prime meridian in the national capital to ensure that the new republic would be scientifically as well as politically independent of all other nations.[106]

Not long after the return of the Lewis and Clark expedition, Jefferson began to plan the establishment of a survey of the American coasts. Several prominent figures have been credited with having originally proposed such a survey, including Albert Gallatin, Robert Patterson, and Ferdinand Rudolph Hassler. Whether or not the original concept can be claimed for Jefferson, he was nonetheless one of its most avid supporters. In 1806 he recommended to Congress the establishment of the survey, a proposal that was favorably received. On February 10, 1807, Congress authorized a survey of coastlines and additional physical features such as islands, shoals, and harbors. Hassler was appointed the first superintendent of the newly formed United States Coast Survey. The survey was not to be just another scientific project, for Jefferson was aware that within the next several years, war with Great Britain was not only possible but probable. The only charts of American coastal waters available to American navigators had been published during the American Revolution and were considerably outdated. As a consequence, it was imperative to undertake a comprehensive survey as quickly as possible, employing the most sophisticated instruments and methods. The coast survey was the last of the major scientific projects to be undertaken during Jefferson's administration, although it was not until long after his retirement that the survey's first fruits were to be realized.[107]

Chapter Eleven

Invention

One new idea leads to another, that to a third, and so on through a course of time until some one, with whom no one of these ideas was original, combines all together, and produces what is justly called a new invention.

—Jefferson to Dr. Benjamin Waterhouse, March 3, 1818[108]

Jefferson's reputation as an inventor was derived primarily from his compulsion to modify and attempt to improve existing utilitarian objects and devices to meet his own requirements, many of which he designed for his own comfort at Monticello. He kept abreast of scientific thought and was keenly aware of the advances being made in the arts and sciences on both sides of the Atlantic. He served, in actuality, as a sort of national information center, collecting news of achievements abroad and relaying them through his network of correspondents to others in the new republic. At the same time he kept his foreign correspondents informed of achievements being made at home. He had a consuming interest in time- and labor-saving, which, combined with his passion for recording statistics related to mundane matters, resulted in the production of numerous devices. Jefferson realized the potential of each new discovery and invention, along with the manner in which it could be adapted for the common good or for his personal convenience. Sophisticated in science and skilled in the practical application of mechanical principles, Jefferson distinguished in technology himself as an inventor in his own right. He had the advantage over many others in his time in that he was an able draftsman, well equipped with both the tools and the skills required to render an idea into detailed tangible form for the guidance of the craftsmen who would produce it. Several of his Monticello slaves were trained with special skills in woodworking and other crafts and could produce what their master designed. Although he owned an elaborate chest of hand tools and always intended to use his own hands to fabricate his devices, he never found time to do so.[109]

Among the devices that appear to have been original with Jefferson was the three-footed or three-legged folding campstool, which he used during his occasional attendance at church services in Charlottesville and while observing the construction in progress of the University of Virginia. In the Dining Room was a dumbwaiter enclosed as part of the fireplace mantel with which wines could be raised from the wine cellar in the basement. This appears to have been a modification of what he had observed in the Café Mécanique in Paris. He also installed mechanized double doors in the Parlor so that both panels moved when one was opened. Another novelty was a folding ladder he designed for use in winding the Great Clock in the Entrance Hall. In the President's House he had a revolving door between the dining room and the pantry that had a set of shelves which "by touching a spring they turned into the room loaded with the dishes placed on them by servants outside the wall." He created a similar arrangement on the door of his Dining Room at Monticello. In his study in the President's House he had designed and installed an unusual map case, built into the wall, in which a number of maps could be rolled up on cylindrical shafts. It was designed in such a manner that any map desired could be rolled out for study, and then when released it rolled back up upon its shaft. Also for his bedchamber in the President's House he designed a clothes closet which had a circular rack pivoted at the center so that when the door was opened the rack could be

Monticello Dining Room (Robert C. Lautman/Thomas Jefferson Foundation, Inc.)

revolved to display the clothing suspended from it. He later incorporated a similar closet in his bedroom at Monticello. Augustus John Foster, who was then secretary of the British legation in Washington, described the closet as "a horse with forty-eight projecting hands on which hung his coats and waistcoats and which he could turn with a long stick, a knick-knack that Jefferson was fond of showing with many other little mechanical inventions."[110]

Revolving Windsor armchair with writing paddle (courtesy of the American Philosophical Society)

Jefferson gave particular attention to any feature that might improve his comfort while reading and writing. For his convenience while writing, he converted a Windsor chair so that it revolved and added a wide panel on one arm for a writing surface. He also designed a combination of a revolving chair and a chaise longue.[111]

While in Philadelphia attending the meetings of the Continental Congress, he arranged with his landlord Benjamin Randolph, who was also a cabinetmaker, to construct a portable lap desk for him based upon his design. He intended to use it primarily for dealing with his correspondence during his long absences from the conveniences of home and during travel, enabling him to utilize what he impatiently described as wasted time during coach travel between Monticello and Philadelphia. The desk was designed with a slanting top that when opened doubled the writing surface; it was also equipped with a book rest to hold reading material. It had a drawer that could be locked, in which he could store documents, pens, ink, blotting sand, and a paper supply. It was to be made of wood sufficiently thin throughout to reduce its weight to a minimum. This was the desk upon which he drafted the Declaration of Independence, and he continued to use it at home and while traveling for the next fifty years. A related convenience of his design was a portable copying press

Portable lapdesk made to Jefferson's specifications and used for drafting the Declaration of Independence (courtesy of the National Museum of American History, Smithsonian Institution)

based upon the invention of James Watt, which he had made for him during his sojourn in England.[112]

"I am not afraid of new inventions or improvements, nor bigoted to the practices of our forefathers," he wrote to Robert Fulton in 1810. "Where a new invention is supported by well-known principles, and promises to be useful, it ought to be tried."[113] Among the most useful of the inventions Jefferson tried was the polygraph, a duplicating writing machine of which he owned at least two examples at the time of his death. In one of his letters, Jefferson described his polygraph "as one of the greatest inventions of the age."[114]

The polygraph was the invention of John Isaac Hawkins, a young Englishman in Philadelphia who was responsible also for the invention of the forte piano and of the physiognotrace, a device for making silhouette portraits. Based upon the principle of the seventeenth-century invention of the pantograph and later perspective machines, the polygraph was a multiple writing machine having two or more pens that simultaneously produced exact copies while the writer used the monitor pen. Hawkins worked closely with his friend Charles Willson

Peale, with whom he shared many interests and skills in the mechanical arts. Hawkins received an American patent for his polygraph in 1803 described as an "Improvement in the pantograph and parallel ruler," and it was patented also in Great Britain later in the same year. Hawkins assigned the American rights to the machine to Peale, who proceeded to manufacture the machine in units having two to five pens. He sold his first unit, one with double pens, to Benjamin Henry Latrobe. Jefferson borrowed it and used it for virtually all his correspondence and other writings that required copies, including his report to the Congress on the state of public buildings in Washington.[115]

Almost immediately Jefferson ordered a polygraph for himself, noting suggestions for modification and improvement. Although he acknowledged that the machine was superior to any other portable copying device, he nonetheless repeatedly requested further improvements. He wished it to be made lighter in overall weight and reduced to the smallest size possible. Jefferson's talents for mechanics were nowhere more evident than in improving the polygraph. He made numerous valuable suggestions for changes in each new model that he obtained from Charles Willson Peale, who valiantly attempted to accommodate the president's

One of Jefferson's polygraphs made by Hawkins & Peale (courtesy of the University of Virginia)

ideas, despite the occasional frustrations he experienced in doing so. With each new modification that Jefferson proposed, Peale did his best to incorporate it and sent the president an altered machine. Jefferson then promptly returned the one he had been using and kept the newer model until he conceived more changes to be made. Many of his suggestions constituted useful improvements of the machine, simplifying its operation. The statesman's mind at work was revealed as he contemplated, experimented, and recommended changes. His emphasis was on ease of use, and his later experiments with reducing the polygraph's size reflected his compulsion for saving effort, space, and time. In the course of the two decades during which he used the polygraph, Jefferson owned at least twelve of them, although because of his exchange program he paid for only several of them. Peale was unable to sell the returned machines, which added to the cost of the operation of his manufactory. At least 139 letters relating to the polygraph and its improvements were exchanged between Jefferson and Peale.[116]

Jefferson proved to be the strongest promoter of the machine. In addition to making gifts of the machine to some of his friends, he used it as "princely presents" or state gifts. Despite the enthusiasm and efforts of Peale, Latrobe, and Jefferson, the polygraph did not live up to its promise, nor did it achieve the popularity that they were firmly convinced it deserved. Although Peale placed advertisements in the press, they did little to increase sales. Then, on the day before Jefferson's re-election to the presidency, he permitted Peale to use in one of his advertisements part of a letter he had written:

> *On five months full tryal of the Polygraph with two pens, I can now conscientiously declare it to be a most precious invention. Its superiority over the copying press is so decided that I have entirely laid that aside. I only regret it had not been invented 30 years sooner, as it would have enabled me to preserve copies of my letters during the war, which to me would have been a consoling possession.*[117]

The polygraph's failure to sell can be attributed to the fact that banking establishments and houses of business were reluctant to replace clerks with equipment that appeared to be extremely intricate. The polygraph's complicated and fragile appearance suggested the need for constant adjustment and repair, which were major deterrents for the general public. Duplicating technology, except for the Watt copying press, was not to come into its own until after the mid-nineteenth century.[118]

In his final years Jefferson maintained two polygraphs, one at Monticello and later another at Poplar Forest, and he continued to use them until a few months before his death. It is due largely to his use of the polygraph that such a large volume of his correspondence and papers has survived.[119]

One of Jefferson's most important inventions, and one that was completely original with him, might have revolutionized diplomatic and military communication from the eighteenth century to modern times if Jefferson had applied it. This was his "wheel cypher," a cryptographic device that employed concentric rings to scramble or unscramble letters in a secret message. While serving as secretary of state between 1790 and 1793, Jefferson became increasingly concerned with the apparent lack of security for confidential communication between his office and representatives of the American government overseas. The war between France and England threatened again and again to bring the United States into the fray, and there were other serious diplomatic problems as well. Jefferson had begun negotiations with the government of Spain relating to his concerns over the navigation of the Mississippi River, a situation that required the most delicate diplomacy. He began to review cryptographic methods that would provide greater security than those in use, and it was almost certainly in this period that he experimented with the wheel cipher.[120]

The American war for independence was responsible in large part for the maintenance of deciphering offices by most of the European nations. Great Britain and France had the best facilities and the greatest degree of expertise, while the colonial Americans were novices in the art of secret communication. Despite the wide range of Jefferson's scholarly attainments, cryptography does not appear to have been among them. During the American Revolution and the period that followed, he seemed to have relied upon the hand conveyance of confidential official correspondence. Now in his present position, however, the amount of diplomatic communication made it essential to have a method that provided the utmost security for correspondence.

After some study, Jefferson devised a method that could be used simultaneously by several correspondents without reducing the security of the system because it was based upon individually assigned keywords. He probably began to experiment with a number of alphabets written in random fashion upon individual strips of paper which he then arranged and rearranged in various ways, no two of them having the letters in the same order. He used twenty-six such strips, each strip numbered from one to twenty-six. The strips were assembled according to the letters of a prearranged numbered key. Generally based upon the numerical value of the letters contained in a phrase mutually agreed upon by both correspondents, the key could be changed periodically upon their agreement. Realizing that the paper strips were too cumbersome to use in the field, even if mounted upon wooden strips, he sought a means to overcome this problem.[121]

English cipher lock, which opens when the word "Rome" is spelled (courtesy of the author)

Unexpectedly, Jefferson found a possible solution upon his office desk in the letter padlocks commonly used in diplomatic service in the eighteenth century for safeguarding diplomatic dispatch boxes. Also known as "cipher locks," they were the precursor of the modern combination lock, and their history can be traced to at least the fifteenth century. Such a lock consisted of four or more brass disks fitted interchangeably upon a central spindle, each disk inscribed along its circumference with a series of letters randomly selected. They were arranged in accordance with a keyword containing as many letters as there were disks, each letter of the keyword appearing on one of the disks. To open the lock, the letters of the keyword had to be aligned. Jefferson also was undoubtedly familiar with the description and illustration of the cipher lock in the *Encyclopédie Ancienne* and *l'Encyclopédie Méthodique*, both of which he owned.[122]

To make the strips less cumbersome in the field, he pasted the strips upon a cylinder that was then segmented into twenty-six disks, each having the letters of

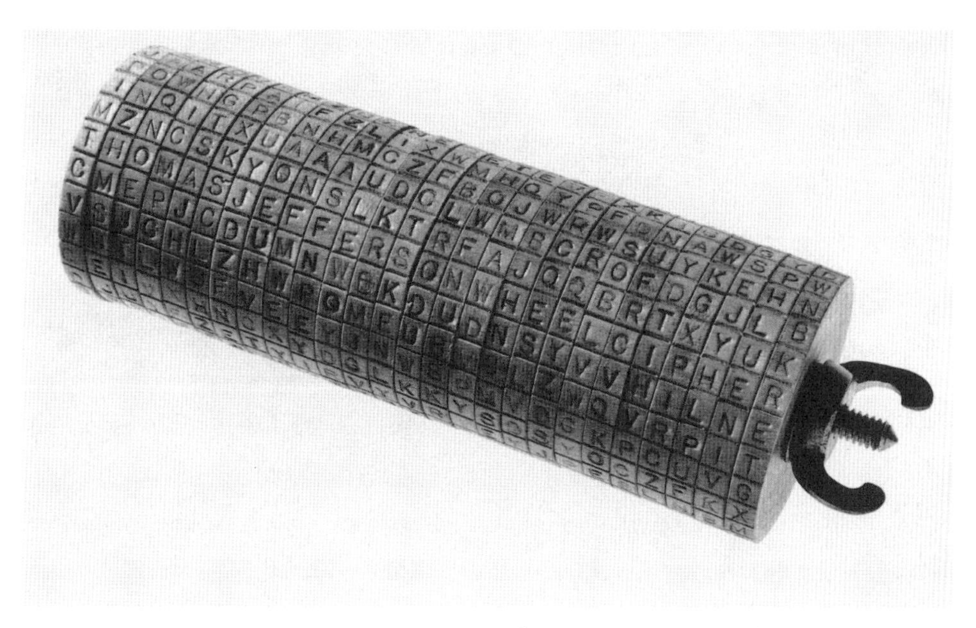

Modern replication of Jefferson's wheel cipher (courtesy of the author)

a random-mixed alphabet inscribed on its periphery. Finding that this arrangement appeared to be operable, Jefferson then had two identical cylinders turned in wood and drilled through the center for insertion of a metal spindle. Next he had each cylinder segmented into twenty-six separate disks. Then he had random-mixed alphabets impressed upon the edges of the disks with steel letter punches. Noted in his Memorandum Books is an entry for a purchase made in April 1792 for a set of punches. The disks were inserted on a spindle and locked into place. The correspondent would line up the disks so that the letters of the agreed-upon keyword were aligned, and then he could use any one of the remaining lines of letters to replace his concealed message. The recipient would receive a message containing the scrambled letters, which he would then line up on his cylinder. Finally, he could then look to the other lines of the cylinder to find the concealed message.[123]

To test the device, Jefferson would have kept one of the cylinders and given the other to a trusted associate, probably Robert Patterson, with whom he had been discussing cryptographic methods. The only documentation concerning the use of the wheel cipher was a letter Jefferson had written to Patterson a decade later, in which he commented upon a cryptographic system that Patterson had devised and submitted to then-president Jefferson for his consideration. "I have

Modern replication of Jefferson's wheel cipher, disassembled (courtesy of the author)

thoroughly considered your cypher," Jefferson wrote, "and find it so much more convenient in practice than my wheel cypher, that I am proposing it to the secretary of state for use in his office." As well as can be determined, Jefferson had discarded the wheel cipher and put it aside after having tested it with Patterson, and never returned to it. All that survives among his papers on the subject are two small sheets in his handwriting, one of which is entitled "The wheel cypher" and the other "Project for a cypher," both of which describe in detail how to make the device.[124]

Although Jefferson apparently set his wheel cipher aside, it was reinvented twice again, the first time after almost a century had passed. It was independently invented in about 1890 by Commandant Étienne Bazeries, chief of the cryptographic bureau of the French Ministry of Foreign Affairs. In an effort to replace the official French ciphers as part of his responsibilities, Bazeries devised two new polyalphabetic ciphers employing twenty random mixed alphabets and incorporated them in a mechanical device he described as "a cylindrical cryptograph." It was identical in form to Jefferson's wheel cipher except for the number of alphabets used. In 1901 Bazeries published a description of his device with an illustration. Although Jefferson and Bazeries might have developed the cipher independent of each other, the similarities are too great to be ignored. It is remotely possible that after Jefferson had abandoned plans for using his wheel cipher for government use, he may have described it to one of his friends in France and that Bazeries found a record of it. There is no evidence to suggest this happened.[125]

It was not until the advent of World War I that the wheel cipher was reinvented once more, this time by a Captain Parker Hitt of the United States Infantry. His invention was based to some degree upon the ideas of Commandant Bazeries, as Hitt admitted. He had, in fact, converted the mixed alphabets of the

Bazeries "Cryptograph" into strip form, rendering them into twenty-five long strips of paper, upon each of which a random-mixed alphabet was inscribed continuously. The strips were numbered and then arranged by means of a pre-arranged numerical key and fitted into a holder made for the purpose. Hitt subsequently rendered the alphabet strips on disks turned from apple wood, which were combined to form a cylinder similar to the one that Jefferson had devised. At some time between September 1916 and August 1917, Hitt's proposal was forwarded to Joseph O. Mauborgne, assistant commandant of the Army Signal Corps at Fort Leavenworth.[126]

In due course Mauborgne reported on Hitt's proposal, commenting that the method was too readily deciphered because the letters had not been sufficiently well mixed. According to Mauborgne's own account, in early 1917 he converted Hitt's strips into a cylinder consisting of twenty-five separate disks, each inscribed with a randomly mixed alphabet, so arranged that the fewest possible number of repetitions of pairs of letters would occur. In mixing the letters on the disks, Mauborgne used the phrase "Army of the U. S." on the R disk in which the letter "R" followed the letter "A." He then had two models constructed for him, in the Signal Corps shop, one of which was made of red fiber and the other of hard rubber, each having the same diameter as the rubber patten on Mauborgne's typewriter, so that the paper strips could be joined exactly and in perfect register when pasted on the perimeter of the disks. The cylinder and disks were turned on a precision lathe and the exact spacing of the twenty-six holes in each, "in one of which in each disk was placed a tiny pin which went only halfway into the disk and projected so as to engage in any disk placed next to it." Alignment of all letters followed.[127]

When Mauborgne was transferred to Washington, he brought one of the models with him and had a series of test enciphered messages set up which were sent for testing to other cryptographers in government service. The experts were unable to decipher any of the messages. Mauborgne then proposed that the cipher device be approved for military use, and it was accepted in 1918. It was not until three years later, however, that the government bureaucracy managed to issue specifications for manufacturing the device, which was produced in aluminum alloy, and had been named "Cipher Device M-94 of the U. S. Army."[128]

Ironically, in 1922, the same year that the Cipher Device M-94 went into service for use by the military, Jefferson's handwritten notes of instructions for making his wheel cipher were discovered for the first time among his papers in the Manuscripts Division of The Library of Congress. In addition to the two handwritten sheets describing the device, it was reported that with them was also a drawing captioned "the figure of the cypher wheel," but it was never located.[129]

The military community was astounded by the discovery in the Jefferson papers, realizing that more than a century earlier Jefferson had the ingenuity to invent the cryptographic device that the Army's trained cryptographers had finally developed, and that Jefferson's detailed descriptions had been available among the public papers during all that time. The annual report of the chief signal officer for 1919 noted "they tried to break this cipher but to date have been unsuccessful, notwithstanding the fact they were given twenty-five messages in the same key." Cipher Device M-94 had a long and distinguished career in official United States military communications for at least two decades after its adoption. It was declared obsolete in 1943 but continued to be used for training purposes thereafter through World War II.[130]

It seemed particularly appropriate that because of his position as secretary of state Jefferson became a member of the three-man board empowered to review and grant patents in the name of the United States. The first patent act, which had been introduced during the first session of Congress in 1789, was enacted into law on April 10, 1790. A patent was granted "if they shall deem the invention or discovery sufficiently useful and important." Jefferson took great pride in being a member of the board, but he was constantly aware of the heavy burden of the responsibility involved, especially since he continued to doubt the constitutionality of the patent system in principle. In his opinion, the American system was modeled too closely upon that of the British, and as a consequence, his dislike of the British prejudiced him against the system. Secondly, he was opposed in principle to granting monopolies that might withhold technological progress from individuals and the American public at large. He also quibbled over the period for which patents were issued, a period that was also copied from the British system. Was the standard period of fourteen years, he wondered, adequate in a country so large and so sparsely settled? He believed that under these conditions an idea would spread much more slowly than in a densely populated country like England. Despite his prejudice against monopolies, Jefferson's experience on the patent board succeeded in persuading him that inventors should be rewarded with a limited monopoly in the arts and manufactures if the new nation was to progress. He later admitted that the patent law "has given a spring to invention beyond my conception."[131]

Many of the petitions received by the patent office were trivial, but Jefferson agreed that some proved to be of importance when put into practice. A new patent law enacted in 1793 abolished the patent board and incorporated some of Jefferson's objections to the original act.[132]

Chapter Twelve

Botany

Botany I rank with the most valuable sciences, whether we consider its subjects as furnishing the principal subsistence of life to man and beast, refreshments from our orchards, the adornment of our flower borders, shade and perfume of our groves, materials for our buildings, or mendicants for our bodies.

—Jefferson to Dr. Thomas Cooper, 1814[133]

Jefferson was an inveterate collector of objects and materials relating to the natural world around him, particularly its flora, fauna, and mineralogical structure. Although he never pursued botany as a science, he had been interested in the subject even during his boyhood, and throughout his life he was interested in plant life that was useful and decorative. In particular he constantly sought examples that were suitable to the needs of American agriculture and horticulture. He included in his *Notes on the State of Virginia* a chapter in which he listed 129 plants, shrubs, and trees native to Virginia that were ornamental, useful for fabrication, esculent, and medicinal. In addition to ornamental plants introduced by English colonists, he also noted fruits and vegetables, to which he added maize, tobacco, potato, pumpkin, cymling, and squash. He was the first to identify the pecan, which he called "Paccan, or Illinois nut" (*Carya sp.*)[134]

Jefferson's hand magnifier (Thomas Jefferson Foundation, Inc.)

Thomas Jefferson's Garden Book, first published by the American Philosophical Society in 1944, was begun in 1766 as a diary of Jefferson's gardening and agricultural interests and is a compilation of varied entries from his Memorandum Books relating to planting at his various properties. Arguably one of Jefferson's great legacies, it was continued, with lapses during his absences from home, until 1824. As Edwin Morris Betts, editor of *Thomas Jefferson's Garden Book,* noted, Jefferson "was possessed of a love of nature so intense that his observant eye caught almost every passing change in it. And whatever he saw rarely escaped being recorded. So we know when the first purple hyacinth blooms in the spring, when peas are up, when they blossom and pod, and when they are ready for the table."[135]

Jefferson exchanged plants at home and abroad and was particularly proud of having introduced dry rice and the olive to South Carolina. Although the culture of dry rice never achieved significance and the olive did not thrive in this climate, still he considered them important achievements, ranking with his authorship of the Declaration of Independence. Writing that "the greatest service which can be rendered any country, is to add an useful plant to its culture," he also imported and cultivated the caper (*Capparis spinosa*).[136]

In 1792 the botanist Dr. Benjamin Smith Barton named a genus of plant *Jeffersonia binata* in acknowledgment of Jefferson's contributions to the natural

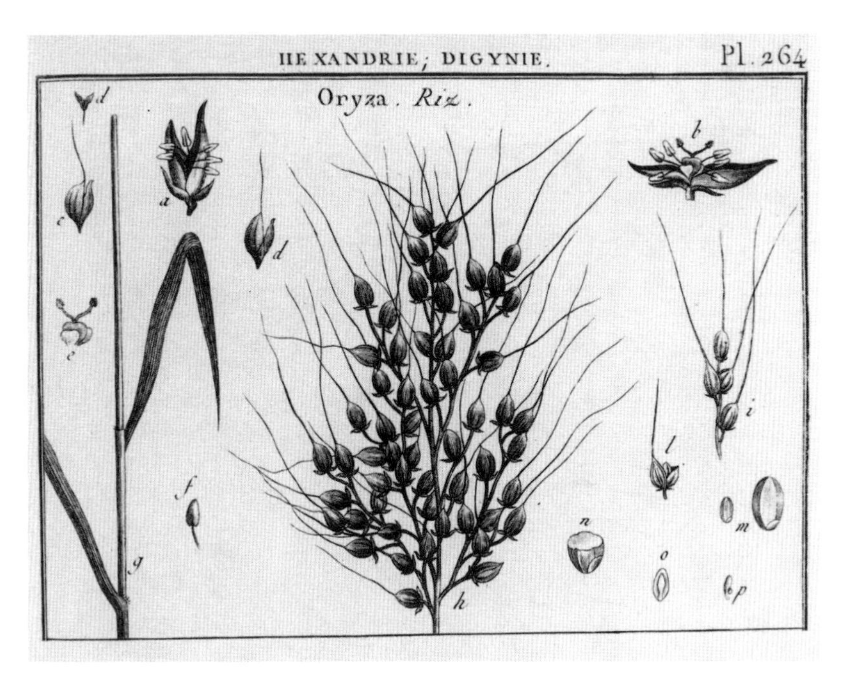

Dry rice introduced by Jefferson to South Carolina

sciences. Later Christian Hendrik Persoon, after Barton, renamed the species *Jeffersonia diphylla*, a name that has remained current. Barton proposed naming the plant in Jefferson's honor because "In the various departments of [natural history], but especially in botany and in zoology, the information of this gentleman is equalled by that of few persons in the United-States."[137]

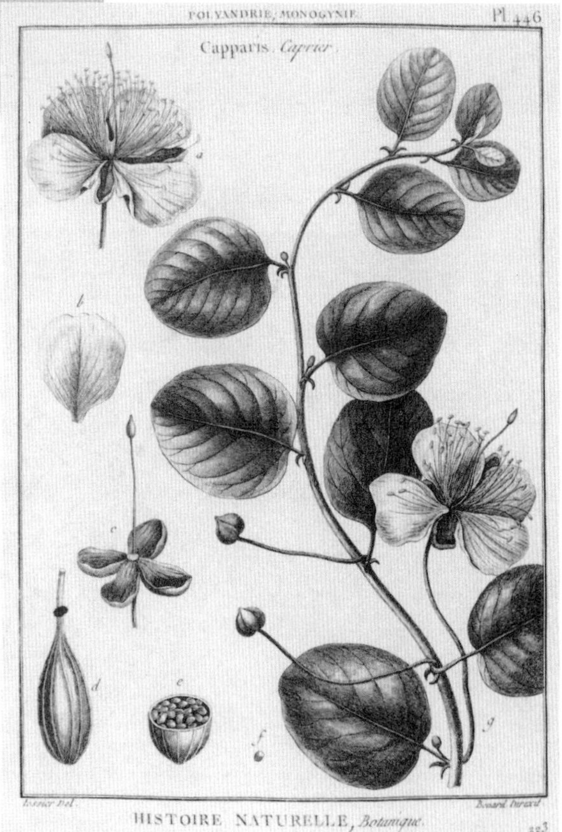

Engraving of Jeffersonia diphylla *(from* Curtis's Botanical Magazine, *1811) and the caper, which Jefferson introduced into the United States after his return from France*

Chapter Thirteen

Horology

Determine never to be idle. No person will have occasion to complain of the want of time who never loses any.

—Jefferson to Martha Jefferson, May 5, 1787[138]

Jefferson's preoccupation with time and the best use to be made of it inspired him during his final year as secretary of state to conceive of a timepiece for Monticello that would serve the entire household within the house as well as those on the farm outside. This was the "Great Clock," which Jefferson designed and which he ordered to be made by a clockmaker of his acquaintance, Robert Leslie of the Philadelphia firm of Leslie & Price. It was designed as a two-faced clock, with one dial in the Entrance Hall to serve the residents, and the other under the East Portico outside the building where farmhands and other workers could see it. Jefferson provided written specifications and measurements of all details, noting that the clock movement "was also to turn an hour hand on the reverse face of the wall on a wooden hour plate of 12 I[nches] radius. There need be no minute hand, as the hour figures will be 6 I[nches] apart. But the interspace should be divided into quarters and 5 minute marks."[139]

Leslie received Jefferson's order just as he was on the verge of departing for England with his family for a protracted stay. Accordingly, he assigned the project to his journeyman clockmaker, Peter Spurck, who soon thereafter established his own clock shop. Some months later, in response to several letters he had received from Leslie, Jefferson informed him, "My great clock could not be made to go by Spruck. I ascribe it to the bungling manner in which he had made it. I was

Great Clock in Monticello's Entrance Hall (Thomas Jefferson Foundation, Inc.)

obliged to let him make the striking mo[ve]ment anew on the common plan, after which it went pretty well."[140]

Jefferson paid Spurck $113.80 for the clock. Although intended for Monticello, the clock had been delivered to the house that Jefferson had rented in Germantown, Pennsylvania, and likely installed there temporarily. When he vacated the premises, he had the clock packed with other furnishings and he brought it back to Monticello. It was not until 1804, however, that Jefferson resolved issues of the clock's installation in the Entrance Hall at Monticello.[141]

The clock movement consists of iron and brass wheelwork encased within a wrought iron frame. The housing is made of wood with a painted chapter ring having the numerals marking the hours and minutes; the seconds are indicated on a smaller dial within the larger one. The clock operates for a period of seven days with a single winding. A pendulum with brass bob swings directly over the front door in the Entrance Hall.

The motive power is provided by a set of fourteen iron falling weights, each consisting of two hemispherical halves assembled together in pairs to resemble cannonballs. They are strung on rope to opposite corners of the front wall and allowed to drop into the basement through openings cut into the floor. The eight cannonball weights on the north wall, or left side of the clock, provide power for the striking mechanism, while the weights on the right side provide power for the time train. Tablets inscribed with the names of the days of the week are attached along the wall alongside one set of weights, so that as they descend, the current day and almost the hour are indicated by the top of the first weight, the weights therefore serving the household as a daily calendar. There was not sufficient space from the ceiling to the floor to include Friday afternoon and Saturday, which are marked on a tablet on the basement wall. On Friday morning the top weight passes into the basement and when the weights have dropped the full length of the rope into the basement, all the weights remain there until Sunday morning when the clock is rewound.[142]

Every Sunday morning Jefferson wound the Great Clock and all other clocks in the mansion. Power for the Great Clock is restored by winding the weights up again around a drum inside the clock housing by means of a special wrought-iron key twenty-two inches long made for the purpose. He designed and had constructed an unusual ladder by means of which the clock could be reached for setting the hands or winding up the weights. Jefferson noted having seen a "folding ladder" in Bergen, Germany, which may have been his source for

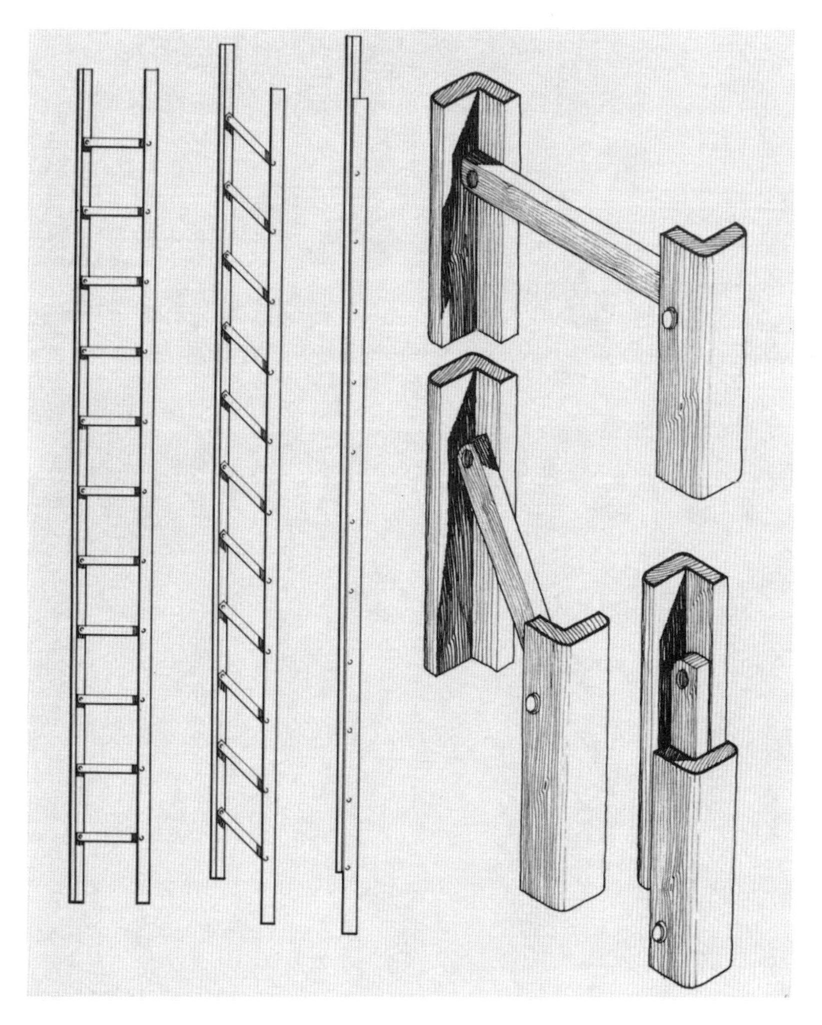

Folding ladder Jefferson designed for winding the Great Clock (Thomas Jefferson Foundation, Inc.)

the design. Or perhaps he was inspired by the draftsman's parallel ruler; the sides of the ladder were made of wooden angle rails with wooden rungs attached on opposite sides of the rails by means of wooden rivets so that they could be pivoted into position to open. By holding the right side and pushing up and over on the left, the ladder is closed with the rails fitted into one another to form a pole. Folded, it remained inconspicuous until it was needed once more, a most ingenious solution.[143]

The Great Clock was intended to activate a sounding device installed on the roof of the building. Jefferson had ordered a Chinese gong "of the shape and size of a camp oven, about 20 I[nches] diameter & 5 I[nches] deep, and weighs about 40 lb. Very coarsely made, being merely hammered out. It therefore can cost little and performs the effect of a very large expensive bell. I wish for one to serve as the bell of the clock, which might be heard all over my farm." The "camp oven" proved to be unsuccessful, however, and had to be replaced.[144]

A gilt ormulu mantle clock on a marble base displayed in the Monticello Parlor (Thomas Jefferson Foundation, Inc.)

Much to Jefferson's disappointment and annoyance, the Great Clock operated unsatisfactorily from the time it was installed, and he ordered some changes to be made in the mechanism. It required repair time and again during the next several years due to the maker's incompetence. Only in recent years, when one of the gears was so badly worn that it had to be replaced, was it discovered that Spurck had made the movement with one tooth short in the gear for winding the weights.[145]

Later in life, Jefferson's preoccupation with time and the most gainful use of it led him to present pocket watches as tokens of esteem to family members or, on occasion, to others. As each of his daughters, and later his granddaughters, reached the age of sixteen, for example, he deemed that she had achieved the status of a young lady with certain responsibilities. On that birthday, he presented each of them with a gold lady's watch. He chose each timepiece carefully to reflect as much as possible the personality of the young recipient.[146]

A passage in *Night Thoughts* by Edward Young, a work that Jefferson had purchased in his youth, which had particularly impressed him, summarized his concern with time to the extent that he copied it into his Literary Commonplace Book:

> *Youth is not rich in time, it may be poor;*
> *Part with it as with money, sparingly; pay*
> *No moment but in purchase of it's worth;*
>
> *And what it's worth, ask death-bells; they can tell.*
> *Part with it as with life, reluctant; big*
> *With holy hope of nobler time to come.*[147]

In his final days, as his physical disabilities increased, Jefferson likened himself to a timepiece, an image that came readily to his mind. As he once commented to his grandson, "I am like an old watch, with a pinion worn out here, and a wheel there, until it can go no longer."[148]

Chapter Fourteen

Agriculture

[Agriculture] is the first in utility, and ought to be the first in respect. The same artificial means which have been used to produce a competition in learning, may be equally successful in restoring agriculture to its primary dignity in the eyes of men.

—Jefferson to David Williams, October 7, 1803[149]

Despite a lifelong career in public service, Jefferson remained a practicing practical farmer. His Farm Book, in which he recorded his farming activities, reveals that he was also a competent agricultural engineer engaged in building roads and fences, as well as working the soil. He was also a horticulturist, experimenting with many varieties of plants and vegetables. He converted Monticello and his other lands into progressive experimental farms for the testing of new crops, agricultural methods, and machines and equipment.

After discovering that his lands were wasting as a result of poor management and neglect during his lengthy absences in his later years, Jefferson was able to restore them to productivity by adopting a seven-year plan of crop rotation and by the addition of gypsum plaster to the soil. His lands had suffered even more by damage caused by the natural elements. His farm's red soil had been washed away by erosion and the recurring planting of the staple crops of tobacco and corn, which contributed not only to the erosion but also the soil exhaustion. Jefferson attempted to counteract these conditions by initiating contour plowing by means of a hillside plow designed by his son-in-law, Thomas Mann Randolph.[150]

Jefferson was among the first to undertake contour plowing to prevent soil erosion. He was particularly proud of the technique of side-hill plowing developed by Randolph, and he wrote lyrically about it to Charles Willson Peale. "The spontaneous energies of the earth are a gift of nature," he wrote, "but they require the labor of man to direct their operation, and the question is, so to husband his labor as to turn the greater quantity of this useful action of the earth to his benefit."

He explained that by plowing horizontally on hilly land following "the curvations of the hills and hollows" on the dead level, every furrow served as a reservoir to receive and retain rain, rather than the water running off and taking the soil with it. Furthermore, the horses drew much more easily on such a level. He noted further that as the declivity of the hill varied in different parts of the line, the guide furrows would approach or recede from each other in different parts of the line and the parallel furrows would finally touch in one part and be far apart in others, leaving unplowed gores, which had to be plowed separately.[151]

On his farm Jefferson made a number of modifications to increase efficiency. Among them was a hemp-brake that could be attached to his sawmill. It was arranged so that it was moved by the gate of his sawmill, which broke and beat at the rate of two hundred pounds a day. As he explained it to Charles Willson Peale in 1815:

> *I have fixed my homony beater differently from yours. I make the saw-gate of my saw-mill move a lever, to the other end of which is suspended a wooden pestle falling into a common homony morter made of a block. All our homony is beaten by this. I make the same saw-gate move another lever at the other end of which is suspended the upper head-block of a common hemp break (but much heavier than common) the break is ranged under the arm of the lever, in the same plane, and the center of it's motion is nearly as may be under that of the lever, while two persons feed the break with the hemp stalks, a third holds the hemp already beaten & formed into a twist, under the head block, which beats it most perfectly; but as one beater is not enough for 2. breakers, I lengthen that arm of the lever 3.f. beyond the point of suspension of the head block, and at the end suspend a pestle, which falling on a block under it, presents a 2d. beater.*

To overcome the inconveniences of such an arrangement, he eventually successfully substituted a horse.[152]

He also arranged a hemp break to be operated with a threshing machine, an arrangement which rendered the procedure simple and inexpensive and "breaks & beats about 80 lb. a day with a single horse. the horizontal horsewheel of the threshing machine drives a wallower and shaft, at the outer end of which shaft is a crank which lifts a common hemp-break the head of which is made heavy enough to break the hemp with it's knives, & to beat it with its head."[153]

Jefferson continued to extol the virtues of farming to Peale, and his enjoyment of it. "I have often thought that if heaven had given me choice of my position and calling," he wrote, "it should have been on a rich spot of earth, well watered, and near a good market for the productions of the garden. No occupation is so delightful to me as the culture of the earth, and no culture comparable to that of the garden. Such a variety of subjects, some one always coming to perfection, the failure of one thing repaired by the success of another, and instead of one harvest a continued one through the year."[154] In 1812 Jefferson cultivated thirty-two varieties of vegetables, twenty-two crops, and thirteen grasses. He exchanged plants and ideas about husbandry with numerous correspondents at home and abroad, and obtained samples of many plants, the cultivation of which he introduced in the United States.[155]

Full-scale model of Jefferson's moldboard plow made by woodworker Robert L. Self and blacksmith Peter Ross (Thomas Jefferson Foundation, Inc.)

Jefferson's promotion of farming methods and conservation practices, his publicizing of new developments and practices that were being made, and his sponsorship of progressive conservation helped to bring about an agricultural renaissance in the early nineteenth century. He supported the formation of agricultural societies as a means of communicating and promoting good farming practices and their development. He held strong beliefs that agriculture would become the basic activity of the new nation. For the new University of Virginia, he included agriculture in his educational curriculum and proposed that a professorship in agriculture be added to the faculty.[156]

Among Jefferson's principal contributions to agriculture was his design of the moldboard plow, which would "receive the soil after the share has cut under it, to raise it gradually, and to reverse it." Always interested in agricultural implements, and constantly seeking means to achieve greater utility, while in France Jefferson had paid to observe a plow drawn by a windlass without horses or oxen. It utilized a complicated rig that enabled four men to do the work of two horses, which in his opinion "was a poor affair." It was while he was traveling from the Rhine back to Paris that his attention had been again drawn to the plow, while observing French peasants near Nancy tilling the soil with clumsy rudimentary plows drawn by oxen.[157]

He became preoccupied with the possibility of designing a moldboard that would be more effective and that could be reproduced simply. Until his time, the shape of the plow was determined by the ingenuity and skills of the individual who made it, and there was no way to reproduce a successful plow. After having studied the various types of plows being used in Europe as well as in America, in April 1788 Jefferson turned to his drafting table and applied mathematical principles to develop the design of a new and more efficient moldboard, one that would be lighter, more easily constructed, and capable of plowing a furrow deeper than before with much less effort.[158]

Until the end of the eighteenth century, the plow retained its ancient form. In most instances it consisted of a crooked stick having an iron tip that was sometimes attached by means of rawhide. Not many of them were designed to turn a furrow, and generally they were made with rough moldboards; no two curves

Drawings (above and opposite) showing the mathematical derivation of Jefferson's design for a moldboard (courtesy of the National Museum of American History, Smithsonian Institution)

were ever alike. Plows made by blacksmiths were of better design but of limited patterns, capable of turning a furrow on soft ground, but when plowing in hard soil, several men and oxen were required. Smiths occasionally produced improved moldboards, but generally their effects were too complicated to be reproduced by the average farmer.

For the most part, wheelwrights produced plows from traditional patterns suitable to their own localities. The moldboard was hewn from wood with the grain running as nearly as possible along its shape. To this were nailed blades of old iron tools or iron straps, or even worn horseshoes, to prevent it from wearing too rapidly. When a moldboard split or broke while in use, the farmer generally cut a section of a tree, and after studying the grain, cut it into shape with an adze.

Although efforts to modify the moldboard had been made many times in the past, in general they followed the traditional designs from the days of the Saxons and were too complicated for duplication.[159]

Noting that other moldboards "were being copied by the eye, so no two were alike," Jefferson set to work to design one that could be readily copied with exactitude. He developed a series of construction diagrams and concluded, "I have imagined and executed a mould-board which may be mathematically demonstrated to be perfect, as far as perfection depends on mathematical principles, and one great circumstance in its favor is that it may be made by the most bungling carpenter, and cannot possibly vary a hair's breadth in its form, but by gross negligence."[160] Jefferson approached the problem by first consulting the works of the English mathematician William Emerson, seeking a formula that would serve his purpose. Basing his work on a study of geometrical solids, he concluded that if the moldboard's function was to pass through the soil with least friction, the wedge was the optimum form to use. Since the other function of the moldboard was to lift and turn the sod, he concluded it "operated as a transverse or lifting wedge." Jefferson then defined a geometrical shape consisting of a series of wedges operating in several intersecting planes in which "a curved plane will be generated whose characteristic will be a combination of the principle of the wedge in cross sections, & will give what we seek, *the mould board of least resistance*."[161]

When he returned from his sojourn in France, Jefferson discussed the moldboard with his son-in-law and constructed one of his own design, which when tested proved to be eminently successful. He sent it to David Rittenhouse for comments and explained its principle, and the latter responded favorably. With this assurance, he then described the moldboard also to Professor Robert Patterson and to John Taylor, the agriculturist and political writer. He then sought to compare the resistance of his moldboard with that of others, wishing he had a dynamometer such

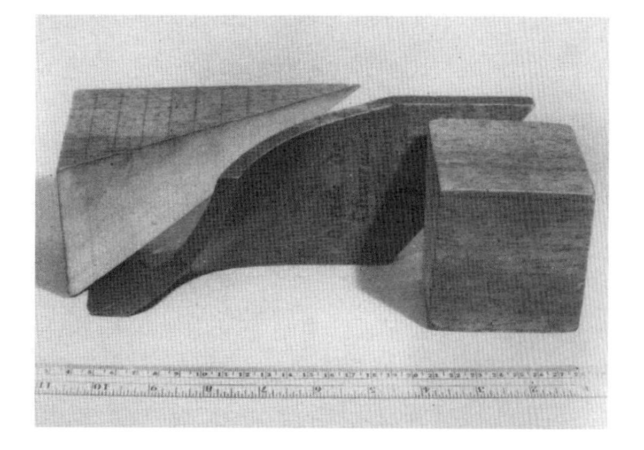

as he had seen used in England, for measuring the force exerted in the drafts of different plows. Dynamometers were hard to find in the United States at that time, but eventually he managed to borrow one and used it to test his moldboard. He discovered that it was too long and that its square toe was liable to collect soil when damp. Accordingly, he modified his original design by shortening it and making it with a sharp toe.[162]

Jefferson produced several models that he distributed to the American Philosophical Society, the Société d'Agriculture de la Seine in Paris, and others. He gave a demonstration of it at Monticello before an English visitor, William Strickland, who presented favorable reports of it to the Board of Agriculture upon his return to England. Jefferson also sent a model with which he enclosed a detailed description and copies of his drawings to Sir John Sinclair, president of the board, noting that he planned to have his moldboard cast in iron. The text of his letter to Sinclair was subsequently published in the *Transactions* of the American Philosophical Society with the title "The description of a mould-board of the least resistance and of the easiest and most certain construction."[163]

Eventually Jefferson had his moldboard cast in iron and fitted to a light wooden plow pulled by two small horses, and used it to till a furrow nine inches wide and six inches deep. His description of the plow was published also in the *Edinburgh Encyclopedia* and in the *Annales du Museum National d'Histoire Naturelle*. The moldboard was displayed at the Philadelphia Society for the Promotion of Agriculture, and in 1807 the Société d'Agriculture de Paris awarded Jefferson a gold medal for his design. Although the moldboard received wide acknowledgment, it did not achieve the universal adoption that had been hoped for it because it was almost immediately supplanted by an all-iron plow developed in the same period.[164]

In 1817 Jefferson became involved with the Albemarle County Agricultural Society that had been organized by a group of his friends and neighbors. He drew up a platform for the new organization captioned "Objects for the Attention and Enquiry of Agriculture." These "Objects" included the production of such staples as wheat, tobacco, and hemp, improvement of the soil, care of livestock, and development of farm machinery. He insisted that members had to file reports of bad practices as well as good ones. As similar societies elsewhere in the state and in other states began to cooperate and exchange publications, the Albemarle County Agricultural Society flourished. On the basis of its success, Jefferson then proposed a "Scheme for a System of Agricultural Societies." In this proposal was the genesis of what later would be established as the Department of Agriculture under the federal government. The connected system of agricultural societies that he had visualized in fact reflects the present system of extension agencies of the state Department of Agriculture and the affiliated Grange societies.[165]

Chapter Fifteen

The University of Virginia

I am for encouraging the progress of science in all its branches; and not for raising a hue and cry against the sacred name of philosophy; for aweing the human mind by stories of raw-head and bloody bones to a distrust of its own vision, and to repose implicitly on that of others; to go backwards instead of forward to look for improvement; to believe that government, religion, mortality, and every other science were in the highest perfection in the ages of the darkest ignorance, and that nothing can ever be devised more perfect than what was established by our forefathers.

—Jefferson to Elbridge Gerry, January 26, 1799[166]

Jefferson first began to contemplate the establishment of a new institution of learning for the state of Virginia as early as 1779. He visualized a university

in which all the branches of science useful to us, and at this day, should be taught in their highest degree, and that this institution should be incorporated with the College and funds of William and Mary. But what are the sciences useful to us, and at this day thought useful to anybody? A glance over Bacon's arbor scientiae *will show the foundation for this question, and how many of his ramifications of science is now lopped off as nugatory.*

He asked others concerned with education, such as Dr. Thomas Cooper, for suggestions of the sciences considered to be the most useful.[167]

From about 1819 Jefferson's attention was directed for the most part to the realization of his dream. After Albemarle Academy was revived and converted to Central College in 1816 by an act of Assembly, a bill for the establishment of a state university was passed and funds were appropriated, with Jefferson as chairman. As plans developed, the statesman's designs were accepted, and he was elected first to the Board of Visitors, and then named Rector of the new university, a position that enabled him to supervise the construction, choose the faculty, select the contents of its library, and prescribe the courses of study. He became, in effect, the self-appointed architect of the university at every level. He not only chose the site on which it was to be built, but also prepared the deed for the land, and he personally laid out the grounds.

Taking with him Monticello's master builder and overseer, he surveyed the site with their assistance. Jefferson provided ten able-bodied men from Monticello's farm staff, supplied the brick, and supervised the construction. While preparing his design for the university, he consulted William Thornton and Benjamin Henry Latrobe, both of whom made suggestions. President James

The University of Virginia Rotunda shown in a lithograph by P. S. Duval (courtesy of Special Collections, University of Virginia)

Monroe laid the cornerstone for what is now Pavilion VII on October 6, 1817, and Jefferson rode from Monticello to the site to observe the progress of the construction every day thereafter except during inclement weather.[168]

The university provided Jefferson at last with an opportunity to design not only the observatory he had failed to build at Monticello, but also an unusual teaching dome. Reflecting his strong lifelong interest in astronomy, he found the subject to be beneficial to mankind on three levels. First of all, he claimed, it provided a means of understanding the architecture of the universe, which in his time was fundamental to the study of natural philosophy. Secondly, observational astronomy was a pleasurable vocation when facilities permitted, and finally, and most importantly, astronomy provided the means for achieving greater precision in surveying and mapmaking required not only for the definition of personal property, but also of national boundaries. He was one of a handful of men in his time who combined the understanding of the theoretical with a competent practical knowledge of the subject, along with an awareness of its potential applications.

Jefferson made an intensive study of the principal astronomical observatories then existing throughout the world for the purpose of identifying and selecting their best features to be incorporated in his design for the university's astronomical observatory. He produced elaborate plans for the new structure, but despite the considerable effort he had expended on the subject, the observatory for the university was not to materialize within his lifetime, chiefly due to a lack of funding.[169]

Plans were developed for a two-year medical school, although there was no provision made for a hospital. It was the first full-time, state-supported, university-based chair of medicine, for which Jefferson arranged to bring Dr. Robley Dunglison from England to be the university's first and only professor of medicine.[170]

When the Virginia legislature appropriated funds for the purchase of a library for the university, Jefferson compiled his own list of titles to be desired, more than seven thousand in number, consisting of approximately 300 works on history, 370 works on law, 180 books on ecclesiastical history, and some on belles lettres, chiefly Greek and Latin. Among the sciences to be taught were botany, zoology, mineralogy, chemistry, geology, and rural economics (which included agriculture). Assuming that twelve dozen lectures would be presented during an academic year, Jefferson suggested that two dozen each should be devoted to botany, zoology, mineralogy, and geology, and eight dozen to chemistry. Or if mineralogy, geology, and chemistry were to be combined into one, then the student's year was to be divided into two parts, one third to the study of botany and zoology, and two-thirds to chemistry, mineralogy, and zoology.[171]

To the supposition that two-thirds of a year could provide only an inadequate knowledge of chemistry, Jefferson responded "we do not expect our schools to turn out their alumni already enthroned on the pinnacles of their respective sciences; but only so far advanced in each as to be able to pursue them by themselves, and to become Newtons and Laplaces by energies and perseverance to be continued through life."[172]

Jefferson proposed that the least amount of time be devoted to the study of geology. "To learn as far as observation has informed us," he wrote to Dr. John P. Emmet in 1826,

> *the ordinary arrangement of the different strata of minerals in the earth, to know from their habitual collocations and proximities, where we find one mineral, whether another, for which we are seeking, may be expected to be in its neighborhood, is useful. But the dreams about the modes of creation, inquiries whether our globe has been formed by the agency of fire or water, how many millions of years it has cost Vulcan or Neptune to produce what the fiat of the Creator would effect by a single act of will, is too idle to be worth a single hour of any man's life.*[173]

The statesman had strong feelings also about metaphysics. In 1819 he wrote to Ezra Stiles, "The science of the human mind is curious, but is one on which I have not indulged myself in much speculation. The times in which I have lived, and the scenes in which I have been engaged, have required me to keep the mind too much in action to have leisure to study minutely its laws of action."[174]

There was little that escaped Jefferson's attention in the planning for the university. He proposed that manual exercises, military maneuvers, and tactics

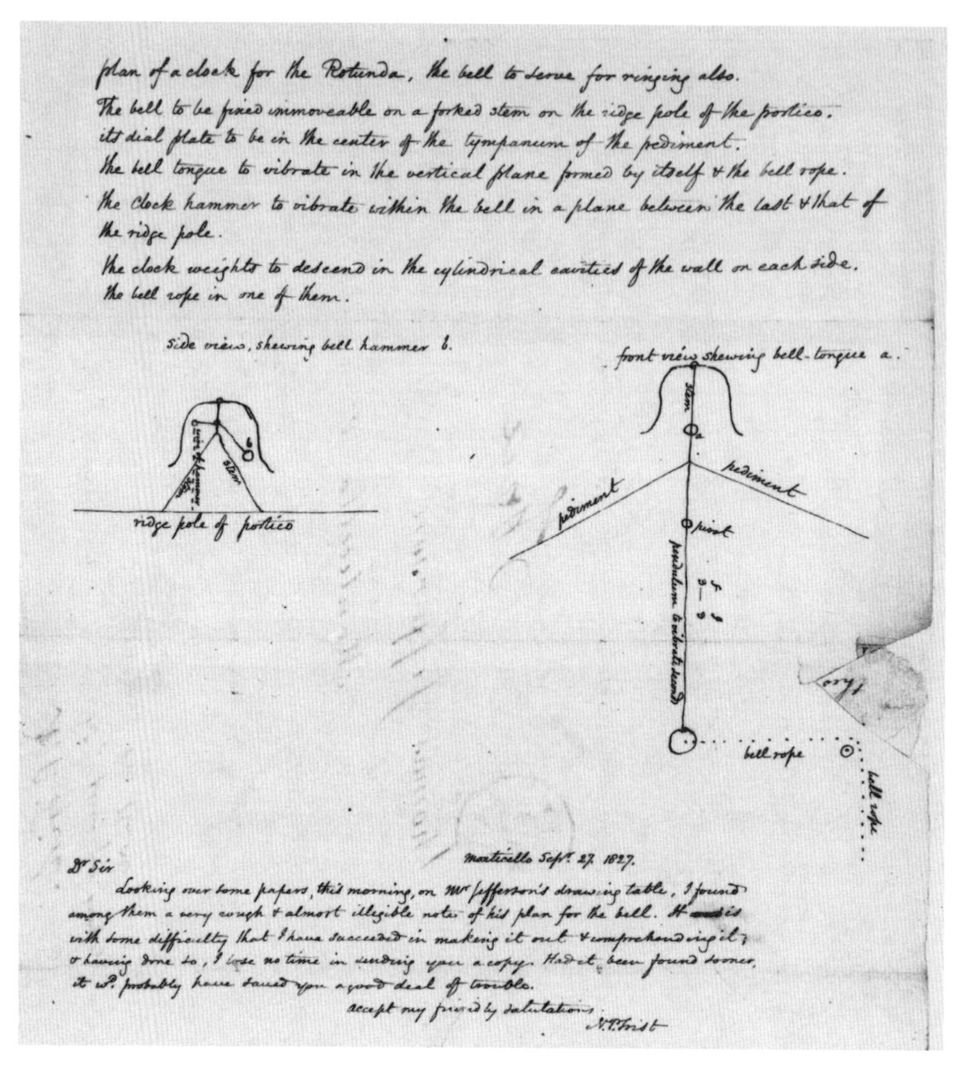

plan of a clock for the Rotunda, the bell to serve for ringing also.
The bell to be fixed immoveable on a forked stem on the ridge pole of the portico.
its dial plate to be in the center of the tympanum of the pediment.
the bell tongue to vibrate in the vertical plane formed by itself & the bell rope.
the clock hammer to vibrate within the bell in a plane between the last & that of the ridge pole.
the clock weights to descend in the cylindrical cavities of the wall on each side.
the bell rope in one of them.

Monticello Sept. 27. 1827.

Dr Sir

Looking over some papers, this morning, on Mr Jefferson's drawing table, I found among them a very rough & almost illegible note of his plan for the bell. It is with some difficulty that I have succeeded in making it out & comprehending it; & having done so, I lose no time in sending you a copy. Had it been found sooner, it wd. probably have saved you a good deal of trouble.

accept my friendly salutations.

N. P. Trist

Jefferson's sketches for the hanging of the bell of the clock for the Rotunda at the University of Virginia (courtesy of Special Collections, University of Virginia)

generally should be the frequent exercises of the students in their hours of recreation. "It is at that age of aptness, docility, and emulation of the practices of manhood," he stated, "that such things are soonest learned and longest remembered."[175] Since the university had not yet established a style of student diet because the boarding houses had not yet been completed, Jefferson proposed a menu that he believed would meet with the board's approval. "Their drink at all times water," he specified, "a young stomach needing no stimulating drinks, and the habit of using them being dangerous."[176]

The planning and realization of the university was to be Jefferson's last endeavor. In view of his lifelong preoccupation with time and the use made of it, it seems particularly appropriate that among his last projects for the university was its public clock. Early in 1825, as the construction of the university was progressing, a member of the Board of Visitors had been attempting, unsuccessfully, to find a maker of public clocks to produce one to be installed in the university's Rotunda. Jefferson took over the project and enlisted the assistance of his grandson-in-law Joseph Coolidge, Jr., who lived in Boston. Jefferson prepared detailed specifications for the timepiece; it required an eight-day movement, a dial plate seven feet in diameter, and a fifteen-hundred-pound bell. "We want a bell," he wrote, "that can GENERALLY be heard a distance of two miles. Because this will ensure its being ALWAYS heard at Charlottesville."[177]

Several months passed before Coolidge responded with "the name of the best clockmaker of this place," Simon Willard. He reminded Jefferson that while he was secretary of state he had awarded Willard several patents for horological inventions. Willard submitted an estimate of $800 for the Rotunda clock, which appeared to be reasonable and was accepted. Arrangements were concluded by Coolidge acting as intermediary for Jefferson, and the contract was finalized in June 1826, with the clock to be ready in September. Jefferson received the letter of agreement three days before his death and did not live to see the clock in place. Willard later commented that Jefferson's plan and specifications for the clock and bell were the only ones he had received during his long career that had been so well designed that when the time came for installation, everything fit to the sixteenth of an inch.[178]

Jefferson was already permanently confined to his bed when he received an invitation to attend the celebration in Washington of the fiftieth anniversary of the signing of the Declaration of Independence. In declining, he carefully worded what was to be his last public statement in which science was featured. As he wrote in part in his response to General Roger Weightman, "All eyes are opened, or opening, to the rights of man. The general spread of the light of science has already laid open to every view the palpable truth that the mass of mankind has not been born with saddles on their backs, nor a favored few booted and spurred, ready to ride them legitimately, by the grace of god. For ourselves let the annual return of this day, for ever refresh our recollections of these rights, and an undiminished devotion to them."[179]

Although Jefferson read widely, he was unable to keep abreast of all worldwide scientific and technological accomplishments, or to give them the attention of a specialist. Occasionally he made erroneous evaluations of the work of others, particularly when his nationalistic bias caused him to react without adequate reflection. He ventured the opinion, for example, that Sir William Hershel's merit "was that of a good optician only," and he was critical of the work of Jan Ingenhousz on plant life and his assertion that light promoted vegetation.[180]

At the same time that this ever-alert and inquisitive mind searched for truth in every dimension and direction, Jefferson's imagination occasionally led him to reach conclusions too quickly. From time to time his momentary enthusiasms caused him to forget his basic precept that the only legitimate conclusions were those based upon careful observation and experiment. His dedication to another of his precepts—that the success of a new nation lay in its application of science and technology—sometimes caused him to become entrapped by naiveté and consequently fail to give sufficient consideration to the numerous proposals offered to him, particularly while he was examiner of patents.

With a mind that was never still and a constant and insatiable curiosity, Jefferson occasionally proved to be inaccurate and impractical in his scientific pursuits, and some claimed that he was too visionary. Nonetheless, he promoted the sciences at every opportunity with the conviction that the emerging new republic must and should utilize science regardless of the source, in order to make it a true democracy: it was by means of science, he believed, that human happiness was to be achieved. "Science," he wrote, "is more important in a republican than in any other government. And in an infant country like ours, we must depend for improvement on the science of other countries, longer established, possessing better means, and more advanced than we are. To prohibit us from the benefit of foreign light, is to consign us to long darkness."[181]

Jefferson's dedication to science brought him ridicule and vituperation from political opponents. "I was bold in the pursuit of knowledge," he recalled to a correspondent in 1814, "never fearing to follow truth and reason to whatever results they led, and bearding every authority which stood in their way."[182] He was attacked again and again for his scientific preoccupations. Particularly damaging was a stanza from a poem entitled "The Embargo" written by a young boy, William Cullen Bryant, who was to become one of America's prominent poets. The stanza in the poem that is most memorable and offensive, and which has often been quoted, is the following:

> *Go, wretch, resign the presidential chair,*
> *Disclose the secret measures foul or fair,*
> *Go, search with curious eye, for horned frogs,*
> *Mongst the wild wastes of Louisiana bogs;*
> *Or where Ohio rolls his turbid stream,*
> *Dig for huge bones, thy glory and thy theme;*
> *Go scan, Philosophist, thy* ****** *charms,*
> *And sink supinely in her sable arms;*
> *But quit to abler hands the helm of state,*
> *Nor image ruin on thy country's fate.*[183]

However, Jefferson's perception of the importance of science, which in his time also included technology, along with his belief that it should be supported by government, have received greater recognition in modern times than in his own. In our day we can acknowledge the vision of a great American statesman who happened to be one of the country's foremost men of science. His hope for the future was expressed in a letter to Dr. Benjamin Waterhouse, written on March 3, 1818, a few years before his death:

> *When I contemplate the immense advances in science and discoveries in the arts which have been made within the period of my life, I look forward with confidence to equal advances by the present generation, and have no doubt they will consequently be as much wiser than we have been as we than our fathers were and they than the burners of witches.*[184]

Abbreviations

APS	American Philosophical Society
DAB	*Dictionary of American Biography*
DLC	The Library of Congress
Farm Book	*Thomas Jefferson's Farm Book*, edited by Edwin M. Betts
Garden Book	*Thomas Jefferson's Garden Book*, edited by Edwin M. Betts
L&B	*Writings of Thomas Jefferson*, edited by A. A. Lipscomb and A. E. Bergh
Malone	*Jefferson and His Time* by Dumas Malone
MHi	Massachusetts Historical Society
Papers	*The Papers of Thomas Jefferson*, edited by Julian P. Boyd, et al.
PPAmP	American Philosophical Society Library
TJ	Thomas Jefferson as the writer or recipient of correspondence
MB	*Jefferson's Memorandum Books*, edited by James A. Bear, Jr., and Lucia Stanton
Notes	*Notes on the State of Virginia* by Thomas Jefferson
TJF	Thomas Jefferson Foundation (collections, construction files; formerly the Thomas Jefferson Memorial Foundation)

Notes

1 Letter from TJ to Pierre Samuel Du Pont de Nemours, March 2, 1809, L&B, 12: 260.

2 Letter from TJ to Peter Carr, August 19, 1785, *Papers,* 8: 405-6.

3 Harrison, 10; Malone, 1: 22-23; Bedini, *Statesman*, 9-10.

4 Bullock, 39-60; Rawlings, 95-96; Malone, 1: 101-3; L&B, 13: 144-49; TJ to Benjamin Rush, Jan. 16, 1811, L&B, 13: 1-9.

5 Malone, 1: 51-55, 73-74, 77-90, 102, 104; Bedini, *Statesman*, 24-34, 43-44.

6 Malone, 1: 77-78; Bedini, *Statesman*, 25-32, 42-43.

7 Malone, 1: 116-17; Bedini, *Statesman*, 44-45.

8 *Notes*, 276-67, n. 104.

9 Letter from TJ to Louis H. Girardin, March 18, 1814, L&B, 14: 124-26.

10 Malone, 1: 22-27; Bedini, *Statesman*, 7-11.

11 Entry for Jan. 12, 1778, *MB*, 1: 456.

12 *Garden Book,* 80, 84-85; Bedini, *Statesman*, 77-78.

13 College of William and Mary, Archives, *Faculty Minutes*, October 14, 1773, "Commission of Thomas Jefferson as surveyor"; Bedini, *Statesman*, 61.

14 Bedini, *Statesman*, Doc. 2, 61, 500-2; Bedini, *With Compass and Chain*, 305-12; *DAB*, "Thomas Marshall," 6: 328-29.

15 *Papers,* 2: 14-15, 208; Gaines, 23-28; Swem, map 254.

16 Entries for March 21, April 4, 11-12, 15, 22, 26, 1786, *MB*, 1: 614, 618, 620-23; Bedini, *Statesman*, 151.

17 "Astronomical Equatorial Instrument," UK Patent 1112, April 27, 1776; letter from TJ to David Rittenhouse, Aug 12, 1792; "Design for telescope attachment," N.D., Missouri State Historical Society, Invoice from W. & S. Jones, July 23, 1805.

18 Robinson, 2-23.

19 Letter to Benjamin Vaughan, July 23, 1788, *Papers,* 13: 394-98.

20 Entries for Sept. 2, 1791, Jan. 3, 1794, *MB*, 2: 832, 909.

21 Entries for Sept. 2 and Oct. 5, 1791, *MB*, 2: 832, 836.

22 Letter from TJ to James Clarke, May 22, 1807, DLC.

23 Bedini, *Statesman,* 375.

24 Letter from TJ to Thomas Cooper, Oct. 27, 1808, L&B, 12: 180-82.

25 "Autobiography," in Padover, 1153.

26 Letter to TJ from James Clarke, July 6, 1817, DLC; letter from TJ to James Clarke, Aug. 1, 1817, DLC.

27 Letter to TJ from James Clarke, July 6, 1817, DLC.

28 Ibid.

29 Letter from TJ to James Clarke, Aug. 1, 1817, DLC.

30 Letter to TJ from James Clarke, Sept. 1, 1818, DLC.

31 Letters from TJ to James Clarke, Sept. 5, 1820, Aug. 15, 1821, DLC; Morrison, 152.

32 Letter from TJ to James Clarke, January 19, 1821, DLC; Clark to TJ, April 2, 1821.

33 Letter to TJ from James Deneale, July 2, 1820, DLC; letter from TJ to Deneale, July 8, 1820, DLC.

34 In the collection of Thomas Jefferson Foundation; there is no mention of its purchase in the *MB*.

35 Letter from TJ to George F. Hopkins, Sept. 5, 1822, L&B, 15: 394-95.

36 Entry for July 4, 1776, *MB*, 1: 420; Randall, 1: 179; Bedini, *Desk*, 15.

37 Weather Memorandum Book, DLC, entry for Sept. 15, 1776; [Nettleton], 308-12.

38 Letter from TJ to Isaac Zane, November 8, 1783, *Papers*, 6: 347-49.

39 Letter to TJ from the Rev. James Madison, Jan. 22, 1784, *Papers*, 6: 507-8.

40 Letter from TJ to James Madison, March 16, 1784, *Papers*, 7: 30-32; letter to TJ from James Madison, April 25, 1784, *Papers*, 7: 122

41 Letter from TJ to Payne Todd, Aug. 16, 1816, Missouri State Historical Society.

42 Letter to TJ from the Reverend Madison, Jan. 22, 1784, *Papers*, 6: 507-8; letter from TJ to the Reverend Madison, April 28, 1784, *Papers*, 7: 133-34; letter from Reverend Madison to TJ, April 18, 1784.

43 Letter from TJ to Jacob Bigelow, April 11, 1818, L&B, 19: 259-61.

44 Bedini, *Statesman*, 395.

45 Varro, *De res rusticus*; Castell, *Villas*; Bear, *Jefferson at Monticello*, 12, 125 fn. 38, fn. 41; Sowerby, "Library," p. 87; letter to TJ from Jacob Rubsamen, Dec. 1, 1780, *Papers*, 4:174.

46 *Notes*, 73, 278 n. 1.

47 Letter from TJ to Governor Wilson C. Nicholas, April 19, 1816, L&B, 14: 471-87.

48 Letter from TJ to Rittenhouse, July 19, 1778, *Papers*, 2: 202-3.

49 Letters from TJ to William Lambert, Sept. 19, 1811, Dec. 29, 1811, DLC; letters to TJ from Lambert, Nov. 14 and 22, 1811, DLC.

50 Letter to TJ from the Reverend Madison, Nov. 19, 1811, DLC; letter from TJ to the Reverend Madison, Dec. 28, 1811, L&B, 19: 183-84.

51 Letter from TJ to Payne Todd, Oct. 10, 1811, L&B, 13: 94-95; letter from TJ to Lambert, Dec. 29, 1811, DLC.

52 *DAB*, "John William Gerard De Brahm," 3: 182-83; letter from TJ to David Rittenhouse, Aug. 12, 1792, DLC; "Mathematical Apparatus," MHi, *Coolidge-Jefferson Papers*.

53 Letter to TJ from the Reverend Madison, Nov. 19, 1811, DLC; letter from TJ to the Reverend Madison, Dec. 28, 1811, L&B, 19: 183-84; Nichols, *Architectural Drawings*, N-172 (K154a), MHi; TJF Construction File.

54 Letter from TJ to David Rittenhouse, July 19, 1778, *Papers*, 2: 202-3; letter from TJ to Robert Patterson, Sept. 11, 1811, L&B, 13: 85-89; letter to TJ from Patterson, Nov. 11, 1811, PPAmP, *Patterson Papers*; letter from TJ to Patterson, Nov. 10, 1811, L&B, 13: 95-112; letter to TJ from Patterson, Nov. 30, 1812, DLC; letter from TJ to Patterson Dec. 27, 1812, PPAmP, *Patterson Papers*; letter to TJ from Patterson, Jan. 12, 1813, DLC; letter from TJ to Thomas Voight, April 9, 1813, PPAmP; letter to TJ from Patterson, April 24, 1813, DLC; letter from TJ to Patterson, April 9, 1813, PPAmP, *Patterson Papers*; letter to TJ from Patterson, Sept. 3, 1813,

PPAmP, *Patterson Papers*; letter to TJ from Patterson, Aug. 25, 1815, PPAmP, letter from TJ to Patterson, Oct. 13, 1815, PPAmP, *Patterson Papers*; Bedini, *Statesman*, 516-17.

55 Letter from TJ to John Hollins, February 19, 1809, L&B, 12: 252-53.

56 Letter to TJ from the American Philosophical Society, Jan. 7, 1797, APS *Transactions*, 4 (1799): xi-xii; letter to TJ from Benjamin Rush, Jan. 4, 1797, DLC; Randall, 1: 154. See Gilbert Chinard's "Thomas Jefferson and the American Philosophical Society," in *Thomas Jefferson and the Sciences*, I. Bernard Cohen, ed., (New York: 1980).

57 Phillips, 193, 195; "Resolution," Apr. 17, 1792, APS, *Benjamin Smith Barton Papers*; *General Advertiser*, Aug. 23, 1792; TJ to Charles Thomson, *Papers*, 20: 244-45.

58 "Letter to Sir John Sinclair, concerning a description of the mould-board of the least resistance, and of the easiest and most certain construction," APS *Transactions*, 4 (1799): 313-22; "A Memoir on the Discovery of Certain Bones of an animal of the clawed kind in the Western parts of Virginia, APS *Transactions*, 4 (1799): 246-60. See also Chinard, "Thomas Jefferson and the American Philosophical Society," (see note 56, above) and Boyd, "The Megalonyx, the Megatherium and Thomas Jefferson's Lapse of Memory," also reprinted in *Thomas Jefferson and the Sciences.*

59 Letter from TJ to John Adams, Oct. 28, 1813, Capon, *The Adams-Jefferson Letters*, 387-92.

60 Letter from TJ to James Madison, Nov. 11, 1784, *Papers*, 7: 503-7; letter from TJ to Charles Thomson, Nov. 11, 1784, *Papers*, 7: 518-19; letter to TJ from James Madison, April 27, 1785, *Papers*, 8: 110-6; Robins, 112-14; letter to TJ from Charles Thomson, March 6, 1785, *Papers*, 8: 15-16; letter from TJ to Thomson, Oct. 8, 1785, *Papers*, 8: 598-99; letter from TJ to John Vaughan, May 15, 1791, *Papers*, 20: 420; letter from TJ to Thomson, Nov. 11, 1784, *Papers*, 7: 518-19; letter to TJ from Thomson, March 6, 1785, *Papers*, 8: 15-16.

61 Letter from TJ to Charles Thomson, April 22, 1786, *Papers*, 9: 400-01; letter from TJ to John Page, May 4, 1786, *Papers*, 9: 444; letter from TJ to Robert Livingston, Feb. 28, 1799, L&B, 10: 117-18; letter from TJ to the Reverend Madison, Oct. 2, 1785, *Papers*, 8: 574-76.

62 Letter from TJ to Charles Bellini, Sept. 30, 1785, *Papers*, 8: 568-70.

63 Letter from TJ to Thomas Mann Randolph, Jr., Nov. 28, 1796, L&B, 18: 200; "Memoranda," in Padover, 799; McLaughlin, 163-64, 252-53.

64 Letter from TJ to David Williams, Nov. 14, 1803, L&B, 10: 428-31.

65 "Autobiography," in Padover, 1158-59.

66 *Notes*, 43-58, 269-72.

67 Ibid., 45, 269.

68 Ibid., 55-56.

69 *Notes*, 62; [Webster] 1: 364, 371-72; Jones, 333-48.

70 Bedini, *Statesman*, 131-32.

71 Letter from TJ to James Monroe, May 21, 1784, *Papers*, 7: 280; letter from TJ to Francis Hopkinson, Jan. 13, 1785, *Papers*, 7: 602-3; letter from TJ to James Monroe, Jan. 14, 1784 [1785], *Papers*, 7: 608; letter from TJ to Niel Jamieson, Jan. 3, 1785, *Papers*, 7: 603.

72 Letter from TJ to Alexander Donald, Sept. 17, 1787, *Papers*, 12: 132-34; Fay, 202; Verner, 96-108.

73 Letter from TJ to John Adams, June 11, 1812, Capon, 305-8.

74 Ibid.

[75] Ceram, "First American," 8-9; Ceram, "Jefferson's Dig," 38-41.

[76] Letter from TJ to Levett Harris, April 18, 1806, L&B, 11: 101-3.

[77] Letter from TJ to Dr. John Sibley, May 27, 1805, L&B, 11: 7-8; letter from TJ to Benjamin Smith Barton, Sept. 21, 1809, L&B, 12: 312-14; Phillips, 258; *Notes*, 101-2, 282.

[78] Letter to TJ from Isaac A. Coles, March 13, 1809, DLC; letter from TJ to Benjamin Smith Barton, Sept. 21, 1809, L&B, 12: 312-14; letter from TJ to J. S. Barnes, Aug. 3, 1809, DLC; Randall, 3: 321-22.

[79] Letter from TJ to Dr. John Crawford, Jan. 2, 1812, L&B, 13: 117-19.

[80] La Rochefoucauld-Liancourt, 3: 79-80; Blanton, 60-66.

[81] Letter from Waterhouse to John Adams, Sept. 2, 1800, letter from Adams to Waterhouse, Sept. 10, 1800, Harvard University Countway Library; C. Kahn, 597-609; Hawes, 7-50; *Columbia Centinel*, March 12, 1799.

[82] Letter from Dr. Benjamin Waterhouse to TJ, Dec. 1, 1800, DLC; letter from TJ to Dr. Benjamin Waterhouse, December 25, 1800, MHi; Martin, 19-21.

[83] Ibid.; Bedini, *Statesman,* 311-13.

[84] Letter from TJ to Waterhouse, Sept. 17, 1801, Harvard University Countway Library; letter from TJ to Dr. Shore, Sept. 12, 1891, DLC; letter from TJ to John Vaughan, Nov. 5, 1801, DLC; letter to TJ from Coxe, Jan. 15, 1803, DLC; letter from TJ to Coxe, April 30, 1802, College of Physicians; Coxe, *Practical Considerations.*

[85] Letter from Waterhouse to Dr. Edward Jenner, April 8, 1802, Halsey, 55-56; *Columbia Centinel,* July 11 and Sept. 26, 1801; *New-England Palladium,* July 7, 1801, and Feb. 26, 1802; *Independent Chronicle,* May 20, 1802; Blake, 42; *DAB,* 6: 300; Waterhouse, 37-38. See also Levi Lincoln to TJ, April 17, 1803 in Jackson, *Letters,* 1: 34-36, and TJ to Lewis, June 20, 1803, in Jackson, *Letters,* 1: 61-66.

[86] *Notes,* 44.

[87] Letters to TJ from William Goforth, Dec. 1, 1806, and Jan. 23, 1807, APS; letters from TJ to Caspar Wistar, Feb. 25, 1807, L&B, 11: 158-59; letter from TJ to William Clark, Dec. 19, 1807, L&B, 11: 404-5.

[88] Stein, 398 and 398, n20.

[89] Wistar, *Letterbook* 1794-1817, PPAmP; letter from TJ to Lacépède, July 14, 1808, Jackson, *Letters,* 2: 442-43; Rice, 597-627.

[90] Jefferson, "Memoir on the Megalonyx," *Papers,* 29: 291-99; Boyd, "The Megalonyx,," 433-34; Wistar, "Description of the Bones," 526-31.

[91] Charles Sellers, 57-58; letter to TJ from G. C. Delacoste, April 10, 1807, DLC; letter from TJ to Delacoste, May 24, 1807, L&B, 11: 206-8.

[92] Letters to TJ from William Clark, Oct. 10 and Nov. 10, 1807, DLC; letters from TJ to Caspar Wistar, William Clark, and George Rogers Clark, Dec. 19, 1807, L&B, 11: 403-5, 12: 15-16; Smith, *First Forty Years,* 385.

[93] Letter from TJ to Lacépède, July 14, 1808, Jackson, *Letters* 2: 442-43; Rice, 597-627.

[94] Wistar, "An Account of Two Heads," APS *Transactions,* 1 (1818): 375-80; see also Rice, "Jefferson's Gift of Fossils to the Museum of Natural History in Paris."

[95] Letter from TJ to Rueben Haines, May 18, 1818, Academy of Natural Sciences, Archives; Bedini, "Jefferson, Man of Science," 20-21.

[96] Letter from TJ to Andrew Ellicott, June 24, 1812, L&B, 19: 185.

97 Jefferson's instructions to Lewis, June 20, 1803, DLC, reprinted in Jackson, *Letters* 1: 61-66.

98 Adelman, 39.

99 Jackson, *Stony Mountains*, 86-97; Bedini, "Scientific Instruments"; letter from TJ to Meriwether Lewis, June 20, 1803, Jackson, *Letters* 1: 61-66; Bedini, *Statesman*, 348-49.

100 Letters from TJ to Dunbar, May 25, 1805 and Jan. 12, 1806, L&B, 10: 74-78; Jackson, *Letters* 1: 245.

101 Jackson, *Pike Journals*, 1: 134-89, 285-86; Jackson, *Stony Mountains*, 246-48; letter to TJ from Henry Dearborn, Feb. 11, 1807, DLC; Chuinard, 4-13. See also "A New Kind of Explorer" in *Army Exploration in the American West*, 1803-1863 by William H. Goetzmann (New Haven: Yale University Press, 1959).

102 Letter from TJ to Robert Patterson, March 22, 1802, DLC; Bedini, *Stone*, 14.

103 Harvey, 42-48.

104 Bedini, *Stone*, 62-64; Heine, 1-27; Olszewski, 15; Twain and Warner, 167.

105 Townsend, 17-18; *Review of Report of the Board of U. S. Engineers on the Foundation of the Washington National Monument by the Washington National Monument Society*, April 10, 1877 (Washington: Gibson Brothers, 1877), 12-13; *The Completion of the Washington National Monument. Annual Report and Documents* 1874-1887 (Washington: U.S. Government Printing Office, 1887?), 48-50. Volume in the collection of the Survey Lodge of the National Capitol Park Service, Office of Mall Operations.

106 Bedini, *Stone*, 126-27.

107 Dupree, 29; G. A. Weber, 1-3; Hassler, *DAB*, 4: 385-86.

108 Letter from TJ to Waterhouse, March 3, 1818, L&B, 15: 162-65.

109 Letter to TJ from Thomas Robinson, March 25, 1786, ViU; entry for April 4, 1786, *MB*, 1: 618; Gaynor, 55-60; Hummel, 27-46.

110 Nichols and Bear, *Monticello*, 26; TJF Construction Files; Bedini, *Statesman*, 328, 265-66; Smith, *First Forty Years*, 385-88; Marie Kimball, 80; TJF Construction File; Smith, *A Winter in Washington*, 38; Richard Beale Davis, ed., *Notes on the United States of America Collected in the Years* 1805-6-7 *and* 11-12 *by Sir Augustus John Foster, Bart.* (San Marino, Calif.: Huntington Library, 1954), 144.

111 Bedini, *Desk*, 6-7; Granquist, 1056-60; Fiske Kimball, 58-60; letter from John Kintzing Kane to president of APS, April 20, 1838, PPAmP; Randolph, 401-4; Wilstach, 110, 134-42.

112 Bedini, *Desk*, 6-15; Randolph, 401-4; Bedini, *Copying Machines*, 9-28, 34.

113 Letter from TJ to Robert Fulton, March 17, 1810, L&B, 12: 380-81.

114 Letter from TJ to Latrobe, July 16, 1817, DLC.

115 Hamlin, 189-214; letter from C. W. Peale to Latrobe, Sept. 24, 1803, PPAmP, *Peale Papers*, Letterbooks.

116 Letter from TJ to Peale, Nov. 12, 1806, Horace Sellers, 308-9; Bedini, *Copying Machines*, x, 130-32, 153-59; letters from TJ to Peale, Feb. 27 and March 1, 1804, Charles Sellers, 141-42; letters to TJ from Peale, March 5 and 13, 1804, DLC; letters from TJ to Peale, March 1 and March 30, 1804, Charles Sellers, 142-43; *MB*, 2: 1128, 1144, 1158, 1195, 1414.

117 *Poulson's American Daily Advertiser*, Dec. 6, 1804; Bedini, *Copying Machines*, 98-99.

118 Bedini, *Copying Machines*, 151-53.

119 Ibid., 181-82, 204.

120 H. P. Johnson, 95-105; Ralph Weber, 118-24; letter to TJ from William Carmichael, June 25, 1785, *Papers,* 8: 251; Bedini, *Statesman*, 233-39.

121 David Kahn, 191-92.

122 Tomlinson, 16-26; Sowerby, *Catalog*, 5: 143-50.

123 Entry for April 21, 1792, *MB,* 2: 868; [P. P. Prime], *Prime Directory*, MS, PPAmP, W. J. Bell, "Addenda," 131-70.

124 Letters from TJ to Patterson, March 22 and April 17, 1802, DLC; letters to TJ from Patterson, Dec. 19, 1801, April 12, April 17, and April 24, 1802, DLC; David Kahn, 194-95; Ralph Weber, 170-77; "*The Wheel cypher*," "*Project of a Cypher*," DLC, f. 22138 and f. 41575.

125 Bazeries, *Chiffres secrets*; David Kahn, 246-50.

126 Memorandum from Col. Parker Hitt to the Director, Army Signal School, December 19, 1914, *Hitt Papers*; David Kahn, 324-25.

127 Letter from Major General Joseph O. Mauborgne. USA Ret. to Colonel William F. Friedman, Aug. 16, 1945, the George C. Marshall Memorial Library, *Friedman Collection*; David Kahn, 325.

128 Signal Corps, U.S. Army, *Specifications No.* 72-26, May 20, 1921, George C. Marshall Memorial Library, *Friedman Collection*; David Kahn, 585.

129 Letter from William F. Friedman to John M. Manly, April 12, 1922, University of Chicago, Joseph Regenstein Library, *Manly Papers.*

130 Chief Signal Officer, U. S. War Department, *Instructions for Cipher Device Type M*-94; Kruh, 20-21; David Kahn, 385.

131 Bedini, *Statesman,* 207; Bedini, "Godfather of American Invention."

132 Letter from TJ to Robert R. Livingston, Feb. 4, 1791, Ford, 5: 276-77.

133 Letter from TJ to Dr. Thomas Cooper, Oct. 7, 1814, DLC; *Garden Book*, 534.

134 *Notes*, 38-43.

135 *Garden Book*, 1766-1824, record of time and motion spent; *Garden Book,* v-vi.

136 Letter from TJ to William Drayton, July 30, 1787, *Papers,* 11: 644-50; Thomas Jefferson, "Memorandum of Services," c. September 1800, DLC, reel 22.

137 Benjamin S. Barton, *Transactions of the American Philosophical Society* (Philadelphia: 1793), 3: 334-47, reprinted in *Garden Book.*

138 Letter from TJ to Martha Jefferson, May 5, 1787, Betts and Bear, 40.

139 Bedini, "Clock Designer," 165-67; TJ, "The Great Clock," undated [1793], *Thomas Jefferson Papers*, DLC, vol. 233, fol. 41588 (shown in Bedini, *Statesman*, Doc. 8).

140 Letter from TJ to Robert Leslie, Dec. 12, 1793, *Thomas Jefferson Papers*, DLC, vol. 95, fol. 16353-54; Bedini, "Clock Designer," 168.

141 Entry for April 27, 1793, *MB*, 2: 893; letter to TJ from Leslie, July 10, 1793, and February 13, 1803, DLC; letter from TJ to Robert Leslie, Dec. 12, 1793, Alderman Library, University of Virginia, *Edgehill Randolph Papers.*

142 Bedini, *Statesman*, 327-28.

143 Ibid., 329; Nichols and Bear, 25-26.

144 TJF Construction File; Nichols and Bear, 26; Bedini, *Statesman*, 329.

145 Nichols and Bear, 26; Bedini, "Clock Designer," 168-69.

[146] Bedini, "TJ and Watches," 38-39; Randolph, 345; Randall, 3: 665-67.

[147] Bedini, *Statesman*, 461; Wilson, 100.

[148] Hall, 220-22.

[149] Letter from TJ to David Williams, Nov. 14, 1803, L&B, 10: 428-31.

[150] *Farm Book*, 48, 50; letter from TJ to Richard Peters, March 6, 1816, DLC.

[151] Letter from TJ to Charles Willson Peale, April 17, 1813, Charles Sellers, 403-5; *Farm Book*, 48, 56.

[152] Letter from TJ to Charles Willson Peale, March 21, 1815, DLC; letter from TJ to George Fleming, December 29, 1815, DLC; *Farm Book*, 251-52; Bedini, *Statesman*, 444-45.

[153] Letter from TJ to Charles Willson Peale, May 8, 1816, DLC; *Farm Book*, 253.

[154] Letter from TJ to Charles Willson Peale, Aug. 20, 1811, L&B, 13: 78-79; *Garden Book*, 461-62.

[155] *Garden Book*, 469-76.

[156] Ibid., 106-7; "Scheme for a System of Agricultural Societies," March 1811, ibid., 640-43; Bedini, *Statesman*, 474-75.

[157] Letter from TJ to Ralph Izard, Sept. 26, 1785, DLC.

[158] *Farm Book*, 47-49; letter from TJ to Robert Patterson, March 31, 1798, DLC; letter from TJ to William Strickland, April 25, 1805, DLC.

[159] Jefferson, "Memoranda," in Padover, 184-85.

[160] Letters from TJ to John Taylor, Dec. 29, 1794, L&B, 18: 199; June 4, 1799, MHi; letters to TJ from Robert Patterson, March 27, 1798, L&B, 10: 15-16; March 29, 1798, MHi; March 30 and March 31, 1798, DLC.

[161] "Mouldboard," letter from TJ to John Taylor, Dec. 29, 1794, in Padover, 997-98.

[162] Letter from TJ to Pierre Samuel Du Pont de Nemours, Feb. 12, 1806, DLC.

[163] "Letter to Sir John Sinclair, containing a description of the mould-board of least resistance, and of the easiest and most certain construction," *Transactions of the American Philosophical Society*, 4 (1799): 213-22.

[164] "D'une oreille de charrue, offrant le moins de résistance possible, et dont l'exécution est aussi facile que certaine. Par M. Jefferson, président des Etats-Unis d'Amérique," *Annales du Museum National d'Histoire Naturelle*, 1: 322-31; Société d'Agriculture de la Seine, Paris; letter from TJ to Augustin-François Silvestre, May 29, 1807, L&B, 10: 212-13.

[165] Jefferson's "Scheme for a System of Agricultural Societies," March 1811, in Padover, 351-55.

[166] Letter from TJ to Elbridge Gerry, Jan. 26, 1799, L&B, 10: 78.

[167] Letter from TJ to Dr. Thomas Cooper, Aug. 25, 1814, L&B, 14: 199-202.

[168] Bedini, *Statesman*, 451-52; Pierson, 31-34; Bruce, 1: 88, 164-72, 209, 236; Honeywell, Appendix, 248-60.

[169] Bedini, *Statesman*, 453-58; Bruce, 1: 52, 270, fig. 7; Lambeth and Manning, 74-78.

[170] Bean, 669-72; Hart, 51-59; Bedini, *Statesman*, 458-59.

[171] Bedini, *Statesman*, 474.

[172] Letter from TJ to Dr. John Emmet, May 2, 1826, L&B, 16: 168-72.

[173] Ibid.

[174] Letter from TJ to Ezra Stiles, 1819, in Washington, 7: 127.

[175] Adams, 94-96.

[176] Letter from TJ to Mr. La Porte, June 4, 1819, Harvard University Library.

[177] Letter from TJ to Joseph Coolidge, Jr., April 12, 1825, with enclosure, Alderman Library, University of Virginia.

[178] Letters to TJ from Joseph Coolidge, Jr., Aug. 5, 1825, and Feb. 27, 1826, *Edgehill-Randolph Papers*, Alderman Library, University of Virginia.

[179] Letter from TJ to Roger C. Weightman, June 24, 1826, DLC.

[180] Letter from TJ to the Reverend James Madison, July 19, 1788, *Papers*, 13: 179-83.

[181] Letter from TJ to an unidentified correspondent, Sept. 28, 1821, L&B, 15: 339.

[182] Letter from TJ to Dr. Thomas Cooper, Feb. 10, 1814, L&B, 14: 85.

[183] Mabbot, 9-15, 23-24.

[184] Letter from TJ to Waterhouse, March 3, 1818, L&B, 15: 162-64.

Bibliography

Adams, Herbert B., *Thomas Jefferson and the University of Virginia*. U. S. Bureau of Education Circular of Information No. 1. (Washington, D.C.: Bureau of Education, 1888).

Adelman, Seymour, "Equipping the Lewis and Clark Expedition," *American Philosophical Society Bulletin for 1945*, (1946), 39-41.

Bazeries, Étienne, *Les chiffres secrets dévoilés* (Paris: Librairie Charpentier et Fasquelle, 1901).

Barton, Benjamin Smith, "A Botanical Description of the PODOPHYLLUM DIPHYLLUM of Linnaeus," *American Philosophical Society, Transactions*, vol. 3, 1793, pp. 342-44.

Bean, William B., "Mr. Jefferson's Influence on American Medical Education," *Virginia Medical Monthly*, vol. 87, December 1960, pp. 669-80.

Bear, James A., Jr., ed., *Jefferson at Monticello* (Charlotesville, Va.: University of Virginia Press, 1967).

Bear, James A., Jr. and Lucia C. Stanton, eds., *Jefferson's Memorandum Account Books: Accounts, with Legal Records and Miscellany, 1767-1826* (Princeton: Princeton University Press, 1997).

Bedini, Silvio A., *Declaration of Independence Desk: Relic of Revolution* (Washington: Smithsonian Institution Press, 1981).

————, "Godfather of American Invention," *The Smithsonian Book of Invention*, edited by Robert C. Post (New York: W. W. Norton, 1979), pp. 82-85.

————, "Jefferson and Science," *Catalogue of Exhibition* for the occasion of the tenth annual Jefferson Lecture of the National Foundation of the Humanities, 1981.

————, "The Jefferson-Coxe Correspondence on the Smallpox," *Fugitive Leaves from the Historical Collections of the College of Physicians of Philadelphia*, 3rd series, vol. 8, No. 1, Spring 1993, pp. 1-4.

————, "Jefferson, Man of Science," *Frontiers, Annual of the American Academy of Sciences of Philadelphia*, vol. 3, 1981-82, pp. 10-23.

————, "Jefferson, Man of Science," in *Thomas Jefferson The Man ... His World, His Influence*, edited by Lally Weymouth (New York: G. P. Putnam & Sons, 1973), pp. 128-38 & illus.

————, *The Jefferson Stone. Demarcation of the First Meridian of the United States* (Frederick, Md.: Professional Surveying Publishing Co., 1999).

————, "Man of Science," in *Thomas Jefferson. A Reference Biography*, edited by Merrill D. Peterson (New York: Charles Scribner's Sons, 1986), pp. 253-76.

————, "Thomas Jefferson," *Dictionary of Scientific Biography* (New York: Charles Scribner's Sons, 1973), pp. 88-90.

————, "Thomas Jefferson and American Vertebrate Paleontology," *Virginia Division of Mineral Resources Publication 61*, Charlottesville, Va., Department of Mines, Minerals and Energy, 1985, pp. 1-26.

————, "Thomas Jefferson, Clock Designer," *Proceedings of the American Philosophical Society*, vol. 108, No. 3, June 1964, pp. 163-80.

————, *Thomas Jefferson and His Copying Machines* (Charlottesville: University Press of Virginia, 1984).

————, "Thomas Jefferson and His Watches," *Hobbies*, February 1957.

————, *Thomas Jefferson Statesman of Science* (New York: Macmillan Publishing Co., 1990).

———, "Thomas Jefferson (1743-1826) Statesman Surveyor," in *With Compass and Chain, Early American Surveyors and Their Instruments* by Silvio A. Bedini (Frederick, Md.: Professional Surveyor Publ. Co., 2001), pp. 649-58.

———, "The Scientific Instruments of the Lewis and Clark Expedition," *Great Plains Quarterly*, Winter 1984, vol. 4, No. 1, pp. 54-69.

Bell, Whitfield J., Jr., "A Box of old Bones: A note on the Identification of the Mastodon, 1766-1806," *Proceedings of the American Philosophical Society*, vol. 93, No., 2, May 1949, pp. 169-77.

Betts, Edwin M., ed., *Thomas Jefferson's Farm Book* (Princeton: Princeton University Press, 1953; reprint, Charlottesville: Thomas Jefferson Memorial Foundation, Inc., 1999).

———, *Thomas Jefferson's Garden Book* (Philadelphia: American Philosophical Society, 1944; reprint, Charlottesville: Thomas Jefferson Memorial Foundation, Inc., 1999).

Betts, Edwin M. and James A. Bear, Jr., *The Family Letters of Thomas Jefferson* (Columbia, Mo.: University of Missouri Press, 1966).

Blake, John B., *Benjamin Waterhouse and the Introduction of Vaccination* (Philadelphia: University of Pennsylvania Press, 1957).

Blanton, Wyndham B., *Medicine in Virginia in the Early Eighteenth Century* (Richmond: Garrett & Massie, 1931).

Julian P. Boyd, et al., eds., *The Papers of Thomas Jefferson* (Princeton: Princeton University Press, 1950--). Twenty-nine volumes to date.

———, "The Megalonyx, the Megatherium and Thomas Jefferson's Lapse of Memory," *Proceedings of the American Philosophical Society*, vol. 102, No. 5, October 20, 1958, pp. 420-35.

Bullock, Helen Duprey, "A Dissertation on Education in the Form of a Letter from James Maury to Robert Jackson, July 17, 1762," *Papers of the Albemarle County Historical Society*, vol. II, pp. 36-60.

Bruce, Philip A., *History of the University of Virginia, 1819-1919. The Lengthened Shadow of One Man.* (New York: The Macmillan Company, 1920).

Cappon, Lester J., ed., *The Adams-Jefferson Letters: The Complete Correspondence Between Thomas Jefferson and Abigail and John Adams* (Chapel Hill: 1988).

Castell, Robert, *The Villas of the Antients* (London: For the author, 1728).

Ceram, C. W., *The First American* (New York: Harcourt Brace & Jovanovich, 1971).

———, "Mr. Jefferson's Dig," *American History Illustrated*, vol. VI, No. 7, November 1971, p. 41.

Chinard, Gilbert, "Thomas Jefferson and the American Philosophical Society," *Proceedings of the American Philosophical Society*, vol. LXXXVII, 1943, pp. 264-65.

Chuinard, E. G., "Thomas Jefferson and the Corps of Discovery. Could He Have Done More?" *The American West*, vol. 12, Nov.-Dec., pp. 4-13.

Coxe, John Redman, *Practical Considerations on Vaccination or Inoculation for the Cow-Pock* (Philadelphia: James Humphreys, 1802).

Dupree, A. Hunter, *Science in the Federal Government. A History of Policies and Activities to 1940* (Cambridge: The Belknap Press, Harvard University Press, 1957).

Ford, Paul Leicester, ed., *The Writings of Thomas Jefferson*. 10 vols. (New York: G. P. Putnam's Sons, 1892-1899).

Gaines, William H., Jr., "An Unpublished Jefferson Map, With a Petition for the Division of Fluvanna from Albemarle County, 1777," *Papers of the Albemarle County Historical Society*, vol. VII, 1946-47 (1948), pp. 23-28.

Gaynor, Jay, "Mr. Hewlett's Tool Chest. Part I," *Chronicle of the Early American Industries Association*, vol. 38, No. 4, December 1985, pp. 57-60.

Granquist, Charles L., "Thomas Jefferson's 'Whirligig' Chairs," *Antiques*, vol. 109, No. 5, May 1976, pp. 1056-60.

Hall, Gordon L., *Mr. Jefferson's Ladies* (Boston: Beacon Press, 1966).

Halsey, Robert H., *How the President, Thomas Jefferson, and Doctor Benjamin Waterhouse Established Vaccination as a Public Health Procedure. (* New York: The author, under the auspices of the New York Academy of Medicine, 1936).

Hamlin, Talbot, *Benjamin Henry Latrobe* (New York: Oxford University Press, 1955).

Hart, Andrew DeJarnette, Jr., "Thomas Jefferson's Influence on the Foundation of Medical Instruction at the University of Virginia," *Annals of Medical History*, 1938, pp. 47-60.

Harrison, Fairfax, "The Northern Neck Maps of 1737-1747," *William and Mary College Quarterly Historical Magazine*, 2nd series, vol. IV, no. 1, January 1924, pp. 1-24.

Harvey, Frederick L., *History of the Washington National Monument and the Washington National Monument Society* (Washington, D.C.: Norman T. Elliott Printing Co., 1902), p. 42-48.

Hawes, Lloyd E., *Benjamin Waterhouse, M.D. First Professor of the Theory and Practice of Physic at Harvard* (Boston: The Francis A. Countway Library of Medicine, 1974).

Heine, Cornelius W., "The Washington City Canal," *Records of the Columbia Historical Society of Washington, D. C., 1953-1956*, vols. 55-56, pp. 1-27.

Honeywell, Roy J., *The Educational Work of Thomas Jefferson*. Harvard Studies in Education vol. 16 (Cambridge: Harvard University Press, 1931).

Hummel, Charles F., "English Tools in America. The Evidence of the Dominys," *Winterthur Portfolio*, vol. 2, 1965, pp. 27-46.

Jackson, Donald, *Thomas Jefferson and the Stony Mountains* (Urbana: University of Illinois Press, 1981), pp. 86-97.

———, *Letters of the Lewis and Clark Expedition with Related Documents, 1783-1854*, 2 vols. (Urbana: University of Illinois Press, 1978).

———, *The Journals of Zebulon Montgomery Pike*. 2 vols. (Norman, Okla.: The University of Oklahoma Press, 1966).

Jefferson, Thomas, *Notes on the State of Virginia* (London, 1787; reprint, with notes, William Peden, ed., Chapel Hill: University of North Carolina Press, 1955), pp. 38-42. (Page references are to reprint edition.)

———, "A Memoir on the Discovery of Certain Bones of a Quadruped of the Clawed Kind in the Western Parts of Virginia," *American Philosophical Society, Transactions*, vol. IV, 1799, pp. 246-322.

———, "The Description of a Mould-Board of the Least Resistance and the Easiest and Most Certain Construction," *American Philosophical Society, Transactions*, vol. IV, 1799, pp. 313-22.

Jones, Anna Clark, "Antlers for Mr. Jefferson," *The New England Quarterly*, 1939, vol. XII, pp. 333-48.

Kahn, C., "History of Smallpox and Its Prevention," *American Journal of Diseases of Children*, vol. 106, No. 597, 1963, pp. 597-609.

Kahn, David, *The Code-Breakers* (New York: The Macmillan Company, 1967), pp. 192-95.

Kimball, Fiske, "Thomas Jefferson's Windsor Chairs," *The Pennsylvania Museum Bulletin*, vol. 21, No. 98, December 1925, pp. 58-60.

Kimball, Marie, "The Epicure of the White House," *Virginia Quarterly Review*, vol. 9, No. 1, January 1933.

La Rochefoucauld-Liancourt, François Alexandre, *Travels Through the United States of North America ... in the Years 1795, 1796 and 1797 ....* 2 vols. (London: R. Phillips, 1800).

Lambeth, William A. and Warren H. Manning, *Thomas Jefferson As an Architect and a Designer of Landscape* (Boston: Houghton Mifflin Company, 1913).

Lipscomb, Andrew A. and Albert E. Bergh, eds., *The Writings of Thomas Jefferson.* 20 vols. (Washington: The Thomas Jefferson Memorial Foundation, 1904).

McLaughlin, Jack. *Jefferson and Monticello: The Biography of a Builder* (New York: Henry Holt & Company, 1988).

Mabbot, T. O., *The Embargo of William Cullen Bryant* (Gainesville, Fla.: Scholarly Facsimiles & Reprints, 1955).

Malone, Dumas, *Jefferson and His Times.* 6 vols. (Boston: Little Brown & Company, 1948-81).

Martin, Henry A., "Jefferson as Vaccinator," *North Carolina Medical Journal*, vol. 7, no. 1 (January), 1881, pp. 1-34.

Morrison, A. J., "Virginia Patents," *William and Mary Quarterly Historical Magazine*, 2nd series, vol. II, No. 3, July 1922, pp. 149-56.

[Nettleton, Thomas], "Observations Concerning the Height of the Barometer, at different Elevations above the Surface of the Earth," *Philosophical Transactions*, vol. 33, 1725, pp. 308-12.

Nichols, Frederick Doveton and James A. Bear, Jr., *Monticello. A Guide Book* (Charlottesville, Va.: Thomas Jefferson Memorial Foundation, 1982).

Olszewski, George G., *A History of the Washington Monument 1844-1968* (Washington: U. S. Dept. of the Interior, April 1971).

Padover, Saul K., ed., *The Complete Jefferson* (New York: Tudor Publishing Company, 1943).

Pierson, Hamilton Wilcox, *Jefferson at Monticello, The Private Life of Thomas Jefferson.* Edited with introduction by James A. Bear, Jr., (Charlottesville, Va.: The University Press of Virginia, [1862] 1967).

Phillips, Henry, Jr., compiler, "Early Proceedings of the American Philosophical Society for the Promotion of Useful Knowledge," *Proceedings of the American Philosophical Society*, vol. 22, No. 119, part 3, July 1885.

Randall, Henry S., *The Life of Thomas Jefferson.* 3 vols. (New York: Derby & Jackson, 1858).

Randolph, Sarah N., *The Domestic Life of Thomas Jefferson* (Charlottesville, Va.: Thomas Jefferson Memorial Foundation, 1978).

Rawlings, Mary, *The Albemarle of Other Days* (Charlottesville, Va.: The Michie Company, 1925).

Rice, Howard C., "Jefferson's Gift of Fossils to the Museum of Natural History in Paris," *American Philosophical Society, Proceedings*, vol. 95, 1951, pp. 597-627.

Robins, F. W., *The Story of the Lamp* (Oxford: Oxford University Press, 1939).

Robinson, Morgan P., "The Burning of the Rotunda," *University of Virginia Magazine*, series 3, October 1905, pp. 2-23.

Sellers, Charles Coleman, *Mr. Peale's Museum, Charles Willson Peale and the First Popular Museum of Natural Science and Art* (New York: W. W. Norton & Co., 1980).

Sellers, Horace W., "Letters of Thomas Jefferson to Charles Willson Peale, 1796-1823," *Pennsylvania Magazine of History and Biography*, vol. 28.

Smith, Margaret Bayard, *A Winter in Washington, or Memoirs of the Seymour Family*. 2 vols. (New York: E. Bliss & E. White, 1824), vol. II, pp. 123-24.

———, *First Forty Years of Washington Society*. Edited by Gaillard Hunt. 2 vols. (New York: Charles Scribner's Sons, 1906).

Stein, Susan R., *Worlds of Thomas Jefferson at Monticello.* (New York: Harry N. Abrams, 1993).

Swem, E. G., "Maps Relating to Virginia in the Virginia State Library and Other Departments of the Commonwealth," *Bulletin of the Virginia State Library*, vol. VII, Nos. 2-3, April-July 1914.

Tomlinson, Charles, ed., *Rudimentary Treatise on the Construction of Locks* (London: John Weale, 1853).

Townsend, George Alfred, *New Washington, or the Renovated City. Reproduced from a Series of Articles in the "Washington Chronicle."* (Washington: Chronicle Publishing Company, 1874).

Twain, Mark and Charles Dudley Warner, *The Gilded Age. A Tale of Today.* (Seattle and London: University of Washington Press, 1959), p. 167.

Varro, Marcus Terentius, *De res rusticus* (Venice: T. Bettinelli, 1728, 1783), Liber III.

Verner, Coolie, "Mr. Jefferson Makes A Map," *Imago Mundi*, vol. XIV, 1959, pp. 96-108.

Washington, Henry A., ed., *The Writings of Thomas Jefferson*. 9 vols. (Washington: Taylor & Maury, 1853-54).

Waterhouse, Benjamin. 1800, 1802. *A Prospect of Exterminating the Smallpox; Being a History of the Variioliae Vaccinae or Kine-pox, Commonly Called the Cox-pox As It Has Appeared in England: With an Account of a Series of Inoculations Performed for the Kine-pox in Massachusetts.* Parts I and II.

Weber, Gustavus A., *The Coast and Geodetic Survey. Its History, Activities and Organization.* Service Monographs of the U. S. Government No. 16. (Baltimore: Johns Hopkins University Press, 1923).

Weber, Ralph E., *United States Diplomatic Codes and Ciphers 1775-1938* (Chicago: Precedent Publishing Company, Inc., 1979).

[Webster, Daniel], *Papers: Correspondence*, edited by Charles M. Wiltse and Harold D. Moser (Hanover, N.H.: University Press of New England, 1974), vol. I.

Wilson, Douglas L., ed., *Jefferson's Literary Commonplace Book* (Princeton: Princeton University Press, 1989).

Wilstach, Paul, *Jefferson and Monticello* (Garden City, N.Y.: Doubleday & Co., 1925).

Wistar, Caspar, "A Description of the Bones Deposited, by the President, in the Museum of the Society, and represented in the Annexed Plates," *Transactions of the American Philosophical Society*, vol. IV, No. 71, 1799, pp. 526-31, pls. X and XI.

———, "An Account of Two Heads Found in the Morass, called the Big Bone Lick, and Presented to the Society by Mr. Jefferson," *American Philosophical Society, Transactions*, vol. 1, New series, 1818, pp. 375-80.

Index

Numbers in italics indicate illustrations.